Computing in Construction

Computing in Construction

Pioneers and the future

Rob Howard MA, RIBA, FIMgt

Butterworth-Heinemann
Linacre House, Jordan Hill, Oxford OX2 8DP
225 Wildwood Avenue, Woburn, MA 01801-2041
A division of Reed Educational and Professional Publishing Ltd

 A member of the Reed Elsevier plc group

OXFORD BOSTON JOHANNESBURG
MELBOURNE NEW DELHI SINGAPORE

First published 1998

British Library Cataloguing in Publication Data
A Catalogue record for this book is available from the British Library

Library of Congress Cataloguing in Publication Data
A Catalogue record for this book is available from the Library of Congress

ISBN 0 7506 3606 8

Composition by Banbury Pre-Press
Printed and bound in Great Britain by MPG Books Ltd, Bodmin, Cornwall

Contents

Preface

The term 'Information Technology' is a convenient one in its abbreviated form, IT, but it is now ready for the next stage in its evolution. Its, or IT's, simple definition is the combination of computing, the term which it superseded, and communications, the most recent agent of change. The abbreviation can cause confusion. I once hurried to Newcastle for a chartered surveyors' regional conference to be in time for a paper entitled 'Europe – The Surveyor's role in IT' which turned out not to be about information technology at all.

I have to admit some complicity in the excessive use of the term IT in conjunction with building. The acronym received official endorsement in 1982, named 'IT Year' by the UK government. A film was produced to celebrate this, by *Building* magazine with support from the Department of the Environment and others, and I acted as technical co-ordinator. It was called 'Building IT' and demonstrated the possibilities of electronic communication between members of the building team. These still remain possibilities today in spite of great advances in hardware, software and communications. The rate at which people and industries change is much slower, and it is they who have restrained the catalytic effect of technology for change. Since then I have also been involved in the 'Building on IT' series of surveys of IT use in construction, and the 'Building IT 2000' and 'Building IT 2005' research reports collecting experts' views on the future. The term has evolved into 'Construct IT', which is now used for the DOE research strategy, a Centre of Excellence based at the University of Salford, and an annual computer exhibition run by the Chartered Institute of Building and the Royal Institute of British Architects.

Perhaps the term IT has now served its time but, as with the technologies covered in this book, it will have to evolve in its own way. With the Internet as the outstanding development in communications, and 'network computer' already trademarked by Oracle, perhaps InterComputing or IC will become the next buzzword. Alternatively, the term Informatics, derived from the French 'informatique' and German 'informatik', would reflect the growing influence of Europe in a single currency of terminology.

This speculation on future terminology is not a prediction, nor are the indications of some future technologies given in this book. It is a recognition of the evolutionary process of IT, which is not always as fast as it might seem, and the contributions of the pioneers and the many who have helped develop information technology into the essential tool it has now become. This book offers guidelines, based on 25 years' experience of evaluating systems for architects, contractors, engineers and surveyors, for coping with the many technologies, all claimed as new and different, which will emerge in future, and assessing their potential value in the design and construction process.

This book can only record the contributions of a few individuals but everyone who has written software, sold systems or provided feedback as a user, has contributed something to the development and application of IT. My own use of computers over 25 years has had little impact on the central development of information technology. I have produced no software but even organizing the exchange of experience between users, analysing levels of system usage and researching the possibilities of future systems, may have had some small impact on the evolution of computing in construction. The most significant contribution has probably been in the field of standards for exchanging data between different CAD systems. The standard, BS 1192 Part 5 [7], was published in 1990 and revised in 1997 [11] after wide implementation, particularly by the Autocad User Group, and a parallel international standard [12], to which I also contributed, was published in the same year. But, even as this was being done, the technology for defining building elements was moving on to object oriented product models containing not only geometry but also attributes and relationships. Few contributions to the development of information technology last for very long but they form part of an evolving process.

I wish to thank all those with whom I have worked since 1971 and a few who were using computers before then, in particular: Boyd Auger, Geoffrey Ashworth, David Taffs and Geoffrey Hutton. Others who have shared my experiences have been colleagues at the Construction Industry Computing Association, without whom I would not have had the time to write this book, and many others, members of CICA and of various institutions, associations and companies. For the production I would like to thank Zoe Youd and her colleagues at Butterworth-Heinemann, and Ed Hoskins and Jaki Howes for their comments. Publication should have been in electronic form, 'in bits rather than atoms', as Nicholas Negroponte says in *Being Digital* [13] but, like him, I am writing for those who may not use information technology currently, but who will need to make decisions about whether or when to adopt it.

Rob Howard

1

Introduction

How many of the successful computer systems of today could have been foreseen ten years ago? Most of them, in terms of their general capabilities, since they are part of an evolutionary process; few in terms of their current stage of development or market share, and even fewer in terms of the firms which dominate the market. Some examples, which have become significant in the building design and construction market, illustrate the difficulty of prediction.

Virtual reality is regarded as a very recent phenomenon yet the concept of interactive visualization of objects modelled within a computer has existed for at least thirty years. The idea of a headset displaying views to the eye which depend upon the direction in which the head is pointing, was demonstrated by Ivan Sutherland in 1968 [1]. It has taken the intervening period for computer processing speeds and efficient software to develop products which are yet to become commercially viable.

The Autocad computer aided design software first appeared in 1982 as a low cost alternative to systems priced at up to £100 000. Its facilities were primitive and, in Release 1, for example, zooming into part of a large drawing took a long time since even the data off the screen was redrawn. Since then it has been greatly enhanced, although the latest version, Release 14, claims a new improvement over its predecessors 'no regens in paper space' which means it is still improving the speed of regenerating drawings on the screen. It now holds a major share of the CAD market by the use of its revenues to add facilities which have now exceeded those of the expensive systems it replaced. This was a marketing phenomenon rather than a technological breakthrough.

Chapter 2: The evolution of technology

History usually records technological developments as a series of events, often credited to famous individuals alone. The reality of development is that many people help to create change. In *The Evolution of Technology* George Basalla [2] identifies four elements in this process: *diversity, continuity, novelty and selection*. Information technology, in its relatively recent history, exhibits all of these with *selection* as a process related as much to markets and fashion as to natural selection. *Diversity* certainly exists in the vast array of new IT products offered, often by small companies hoping to become the next Microsoft. *Continuity* is there but may be hidden by firms which contribute a new product being reluctant to acknowledge its forerunners. Perhaps lawyers should bear some responsibility for products being claimed as totally distinct from others and deliberately coining new terminology to avoid law suits. *Novelty* is everything in IT.

Humans often struggle to catch up with the opportunities information technology offers. At the same time we are trying to understand and simulate the superior workings of the brain. In *City of Bits* [3], Bill Mitchell presents a view of an 'electronic flaneur' of the 1990s, dependent upon the networks and able to communicate from anywhere, perhaps representing several electronic personalities. He envisages intelligent exoskeletal devices communicating sensory perception of all sorts so that it would be possible to participate in virtual rock climbing or to transmit a good night kiss to a child across the world. As an academic and head of the Media Lab at the Massachusetts Institute of Technology, he represents the forefront of current technology, but we can all expect to follow him and to exceed what seems currently possible, if not our wildest imagination.

Towards the information society
The thesis of this book is that, as in most technological development, the use of IT in the construction industry is an evolutionary process and that, if we can understand what has gone before, we will have a better chance of anticipating the future. Although particular individuals and companies are credited with each new breakthrough, many contribute and, while it is difficult to predict who the next computer billionaires might be, there are trends which are continuous and can help us to cope with decisions on technology we will have to make. The evidence for this evolutionary process is presented, both in some of the more fundamental inventions of the past, and in the relatively recent history of computing. Chapter 2 covers these processes and the emergence of the information society. IT, in the form of international communications, has had a profound effect through crossing, and helping to change, political boundaries. It also allows work to be shared around the world between different time zones. Place is less significant than it was and we will all need to establish a mobile electronic persona in future.

Chapter 3: Pioneers of computing in construction
Another facet of technological development is the cyclical nature of invention. Visions are created long before they can be realized. In IT thoughts are focused on the future and past experience is often neglected. Chapter 3 recalls some of the early heroes of computing and its more recent application in design and construction, and recognizes the contributions of their many collaborators. Changes which took place in perception of space and movement created the climate in which a more dynamic

approach to building data could thrive. What would the great masters of modern architecture have achieved with computers? The work of some of the pioneers in US universities and in UK research groups in construction during the 1960s and 1970s is recognized. Wherever possible my own experience, or that of pioneers I have talked to, is referenced by specific examples (and these are presented in boxes to illustrate the general theme). Changes in methods of construction and the measurement and engineering of these have been very influential. Many of the visions described by the pioneers have still to be turned into practical tools or, where this has been done, they may still await wide exploitation.

Chapter 4: Changing conditions in construction
There is a tendency for the very rapid changes in information technology to obscure other, related developments in the fields to which they are applied. The process of building procurement, design and construction has recently been changing faster than ever. This evolution is summarized in Chapter 4 as it provides opportunities for information technology which then acts as a catalyst for further change. The period 1970 to 1995 is summarized as are the recent changes proposed in the Latham report [24] and in other reports on new technology. The roles of building professionals are changing and have been influenced by IT, which also provides tools for them to promote their services and compete with each other.

Chapter 5: Developments in general technologies
Conditions for the success of particular computer applications are examined using general examples but illustrated by their specific uses in construction. The factors which affected their growth are analysed, using my own experience or published statistics, and related to patterns of growth. Such applications as the spreadsheet, or general capabilities like expert systems, are assessed as part of the general march of technology as well as for their particular impact on the construction industry. How do users cope with new technology and what are the patterns in its growth? These patterns are expressed as curves showing the shape of growth overlaid on histograms of the factors that appear to have had most effect on this. The objective is to establish which factors are most influential.

Chapter 6: Successful applications in construction
This chapter concentrates on systems which are aimed specifically at building design and construction. It examines the early preoccupation with automated design software and the success of more routine applications of CAD. It takes stock of progress towards the integrated building model and the human and commercial obstacles to its achievement. It examines the role of standards in providing continuity and offsetting the IT industry's tendency to present similar systems as new and different. Some early projects which were before their time, and their influence on more recent products, are considered. How have the market in construction, and the key applications for

architects, contractors, engineers and surveyors, developed? In each case major factors leading to success or failure are analysed. Views may differ on whether some have yet succeeded or whether others are already doomed to failure, but statistics on growth and usage are presented, largely from surveys carried out by the Construction Industry Computing Association and professional institutions.

Chapter 7: Conditions for success

To analyse the criteria for success in this complex and fast moving field, it is necessary to examine the stages of system development as well as the changes taking place in the construction market. An information technology system can include: hardware, software, communications, data and the application of these to a particular business function. The stages, from concept of a new system or the intention to improve an existing one, through design and marketing, to sales, are elaborated in Chapter 7. Timing is critical to the success of a new system. Factors affecting cost and the benefits to users in construction are analysed to identify the conditions for success. Charting the growth of a well documented technology, such as facsimile transmission during the 1980s, shows typical S-curve growth with market saturation rapidly approached after reaching critical mass (*Figure 1.1*). This also indicated promising growth for Electronic Mail and for Electronic Data Interchange, the more recent progress of which has been less impressive, as shown in *Figure 6.6*. The role of government and industry in

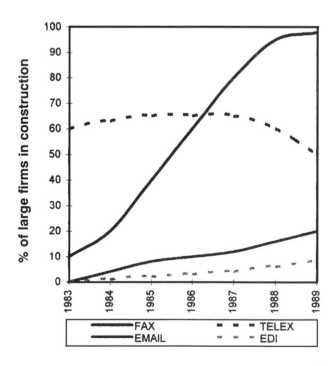

Figure 1.1 S-curve growth of Fax and other communicating technologies. *Building on IT for the 90's.* CICA/KPMG 1990.

research and standards and the time taken to deliver benefits, are all influential on current and future aids to building design and construction. Some technologies were launched before the market was ready for them, but their influence may live on in later products which take advantage of newer technology and a construction industry better educated to use them.

Chapter 8: Some promising developments for the future
The main purpose of this book is to offer some guidelines for assessing future technologies. Much background material from many experts is called upon, not least the Technology Foresight studies of 1995 [4] and Building IT 2005 [5]. The criteria for success or failure, distilled from recent history in Chapters 5 and 6, are applied to the new ideas, presented in such reports, in Chapter 8. Examples are given of less successful projects but only to offer guidelines for assessing future systems. What standards are needed to ensure compatability and integration? These systems will be used by a dispersed and diverse range of people who may, in future, become members of virtual project teams. Promising developments are discussed, many of which have already been proposed or may be future extensions or combinations of existing systems. Some will result from further miniaturization or the combination of processing and communications. Others await changes in the way construction currently organizes itself and a period of change appears to be starting in 1997. Few developments are totally original and demonstrate that all new technology is part of an evolutionary process.

Chapter 9: Lessons for future technologies in construction
The existence of a wonderful new invention does not guarantee that it will be adopted widely. The delivery of useful systems at the most opportune time and with the necessary market conditions for success, is the topic of the final chapter. It discusses the importance of marketing, the opportunities for combining current technologies and the symbiotic relationship between a changing construction industry and new technology. This is as much about future processes in building design, performance analysis, construction and facilities management, as it is about future systems to aid these processes. Information technology is the servant of its users, but getting the most from this very powerful and willing assistant does require some change in the master.

This book can help those appraising new systems to make a better choice which maximizes the lifetime and value of their investment in information technology. It will also help the IT supplier to meet conditions for success in a complex and changing market. Following from the advice offered to both suppliers and users, there should then be a better match between future expectations and the ability to satisfy these. The result will be a more effective and competitive construction industry using appropriate technology as part of a recognized evolutionary process in which the contributions of all are recognized.

2

The evolution of technology – towards the information society

Before addressing the specific application of IT to construction it is worth looking at the nature of technological evolution and the surprisingly long history of computing. The origins of the computer have become a source of dispute between the UK and the USA concerning developments during the 1940s. This period added electronics and general purpose machines to a history going back much further and involving mathematicians from several other countries.

Christopher Evans, in his book *The Making of the Micro* [6], states that computing began when 'man picked up a few small stones or scratched marks in the earth as a kind of record or memory aid'. He then goes on to identify important stages in the elaboration of computing machinery. This records the evolutionary process of which all current and future computer systems and, ultimately all their users, are a part.

The IT and electronics industry is expected to become the biggest in the world by the year 2000. In the UK this market was worth £48 billion in 1993, the same as the turnover of construction, although IT production in the UK was £5 billion less. This market is likely to grow faster then construction, reaching 10% of GDP by 2005. Rates of growth of individual technologies can be phenomenal with traffic on the Internet growing from 1.37 Mbytes/day in 1992 to 52 000 Mbytes/day in 1994 – a factor of 38 000 [4].

The origins of computing go back some way and the development of a broad technology like IT starts quite gradually. The rate of growth is initially limited by the desire and capacity of humans to change, and the social effects of very rapid evolution are an important factor in the general adoption of computing in industry.

2.1 *The way in which technology evolves*

Darwin Among the Machines [8] was written by Samuel Butler in 1863 and takes the view that machines developed in similar fashion to the evolution of living beings. He also predicted a symbiotic relationship between man and machine ultimately leading to self-replication. The evolutionary approach was less popular than the revolutionary, however. Inventions go down in history as the product of one individual or company. George Basalla [2] gives the example of James Watt watching a kettle boil and going on to produce the full sized steam engine, for which he is credited, in 1775. Newcomen's simpler atmospheric steam engines had already existed to pump water out of mines since 1712.

There is a long history of inventions migrating from one industry to another or being of greater use in a different country to that in which they were first produced. The steam engine migrated from the mines to powering mills, railways and steamships. The first digital computer was used to calculate firing tables for the US artillery. Military uses are often behind new products; the Internet was originally Arpanet, a fail safe network for communications.

In medieval times the technologies hailed by Francis Bacon as the greatest inventions of their time were gunpowder, printing and the compass. All were invented in China but exploited elsewhere. It requires an adaptable market to establish the success of a new product. It also requires some perseverance. George Basalla [2] reports some of the limitations of new, and in time, successful inventions when they first appeared: the camera initially required exposure times of several minutes, the first powered flight lasted for only 57 seconds, the first television screens were 4 × 5 centimetres, blurred and flickering, and the first electronic computer occupied 180 square metres of space and weighed 30 tons. It is not surprising that the take up of new inventions is far from assured.

The wheel was known of by the Inca civilization of South America in the first half of this millennium yet they had no wheeled vehicles. Inca toys show the wheel but it was rejected for transport because the terrain in most of their empire was very rugged and they had a good system of trails and relays of messengers. Their llamas carried the loads but were never trained to pull carts. Many years after the invention of gunpowder in the Far East in the ninth century and its introduction to Europe, firearms were reintroduced to Japan by the Portuguese in 1543. They were widely copied and used until the seventeenth century when there was then a rejection of what was regarded as foreign technology in favour of the Samurai tradition of swordsmanship. Gunsmiths returned to making swords and spears until Japan was opened to the West during the nineteenth century and firearms were needed to maintain military balance. Other examples of delay in applying new inventions can be seen in IT, but this is often because their application was dependent upon the emergence of cheaper and better electronics. The virtual reality headset first appearing in 1968 and still awaiting full commercial production is one example. Successful development also depends upon the readiness of markets as the information society emerges from one based on industrialization.

There are plenty of inventions which could clearly never work and the first prototype would have shown this. Perpetual motion has long been sought, from a Sanskrit description of a self-moving wheel in 400–500 AD, to a similar device described by Villard d'Honnecourt in the thirteenth century and, quite recently, Henry

Figure 2.1 Perpetual motion in the seventeenth century. Henry Dirck's perpetuum mobile, from *The Evolution of Technology.*

Dirck's 'perpetuum mobile' published in 1861 (*Figure 2.1*). There are also inventions where their significance may not be fully realized by their inventors. The rights to the transistor were sold, by its American inventor, to Tokyo Telecom in 1953 and they went on to produce miniature radios under their more familiar name of Sony. New types of company may take over the development of a technology. Bill Southwood, in Building IT 2005 [5], gives the example of 'the pocket calculator, which consigned the slide rule to the museum forever, came out of the electronics, not the slide rule, industry'.

So evolution is an erratic process which is deflected or delayed by many factors and we must understand these as well as the conditions for success. A brief look at the key developments in computing up to the 1940s illustrates that the information society did not arrive suddenly. The complete integration of the capabilities of IT with the complexities of a construction industry, which has been evolving for very much longer, is addressed later in this book.

2.2 *A brief history of computing*

Some of the earliest devices for calculation or storage of data were the abacus, which continues in general use in the Far East, notched tally sticks used in medieval Europe and in the British parliament until the eighteenth century, and the knotted quipu used to carry the imperial accounts by runner along the Inca trails in South America. The relatively primitive numerical system of the Greeks and Romans which limited calculation, was enhanced by the Arabs who added the refinement of the zero. From the middle ages the main heroes and their contributions to development of computation are generally recognized as:

- John Napier (1550–1617) Scottish mathematician who invented logarithms. Napier's rods and bones were aids to multiplication.
- William Oughtred (1575–1660) English mathematician who is credited with the invention of the slide rule.
- Blaise Pascal (1623–1662) A Frenchman who produced the first calculating machine which could add and subtract using dials.
- Gottfried Liebniz (1646–1716) A German scientist who perfected the binary system and enhanced Pascal's machine to provide multiplication and division.
- Joseph-Marie Jacquard (1752–1834) A French weaver who introduced punched cards to control weaving patterns.
- Charles Babbage (1792–1871) An English inventor credited as the father of computing for the design of his analytical engine which is regarded as the first general purpose computer. His differential engine was more ambitious and there have been recent attempts to build it (*Figure 2.2*).
- Herman Hollerith, of the USA, used punched cards to reduce the time taken for the 1890 census to one-third of that previously taken. Hollerith's company merged with others to become the Computing, Tabulating & Recording Company, later IBM.
- William Burroughs (1857–1898) An American inventor of the calculating machine went on to form a company in his name which is now part of Unisys.
- Lord Kelvin (1824–1907) A Scottish mathematician and physicist, proposed in 1876 a general purpose calculator, initially to produce tide tables and differential equations. He called it a differential analyser.
- Vannevar Bush, a professor at MIT in the USA, converted this idea into reality in 1930 with a very large mechanical engine driven by electric motor and providing output with pen on graph paper. He went on to propose the Memex in 1945, a device for storing all known information (*Figure 2.3*).

From 1940 onwards computers became electronic and their development was more anonymous owing to military applications which also accelerated their progress.

- Colossus was the code breaking computer developed at Bletchley Park in Britain for breaking the German Enigma codes. It was first operational in December 1943 and employed valves but was dedicated to its particular task.

Figure 2.2 Part of Babbage's differential engine. Source: Science and Society Picture Library.

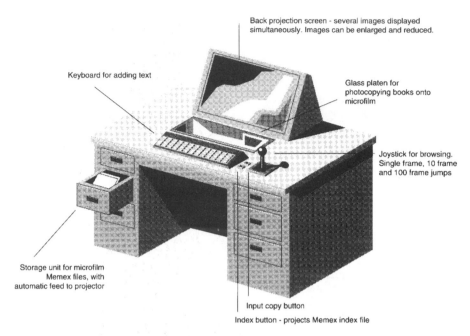

Back projection screen - several images displayed simultaneously. Images can be enlarged and reduced.

Keyboard for adding text

Glass platen for photocopying books onto microfilm

Joystick for browsing. Single frame, 10 frame and 100 frame jumps

Storage unit for microfilm Memex files, with automatic feed to projector

Input copy button

Index button - projects Memex index file

Figure 2.3 The Memex system for access to all known information, 1945, proposed by Vannevar Bush. *Understanding hypermedia.*

- ENIAC, the Electronic Numerical Integrator & Calculator, was completed in the USA in February 1946 and was largely used for calculating artillery trajectories.
- Mark 1, the first computer to run a stored program, Manchester University 1948.
- EDSAC, the Electronic Delay Storage Automatic Calculator, Cambridge University 1949.
- UNIVAC 1 was the first commercially successful computer sold initially to the US Census Board in 1951 and then to 50 other customers.

Subsequent developments evolved from the use of thermionic valves to transistors, integrated circuits and very large scale integration and are well documented. These allowed the miniaturization, speed and reliability, which have turned the computer into a universal machine.

During the 1940s, architects were also thinking about the new architecture which would follow the end of the war and these developments in computing must have had some influence. 'Building for 194X' was a collection of views published in *Architectural Forum* in 1943 [9]. The ideas offered in this by a variety of architects included 'a disposable house' and 'a library stocking nothing but microfilm editions and equipped with electronic brains'.

As the evolutionary process accelerated there were many more stages than can be summarized here, and many contributors whose names are not recorded. From the 1940s developments result from new technologies and companies rather than individual inventors. The most influential were the people who had the bright ideas and often went on to run successful companies: Hewlett and Packard, Jobs and Wozniak (Apple), Bill Gates (Microsoft) and many others. Heroes are still needed for historic purposes. *Byte* magazine's 20th anniversary special issue in September 1995 [10] recognized the contributions of those already mentioned and of Mark Andreeson of Netscape Communications, and Tim Berners-Lee, who developed the World Wide Web, Niklaus Wirth, producer of Pascal and object-oriented languages, and Dennis Ritchie who helped produce UNIX. There are, and will continue to be, names which are singled out although most computing developments are now carried out by large teams and contributed to by many humble users.

2.3 *The importance of digital data*

One influential pioneer, whose background is in architecture, is Nicholas Negroponte, author of *The Architecture Machine* [1] in 1970 and of *Being Digital* [13] in 1995. As director of the Media Lab at the Massachusetts Institute of Technology, he has been at the heart of research into the information society and how data in all media are converging in digital form. Development of IT was originally concerned with hardware and then with software, and now the importance of data has become a priority. Text, graphics, sound and video can be recorded as bits of information which are then combined, processed and communicated electronically.

Negroponte's latest book [13] is in black and white and has no illustrations. This is a surprise coming from someone who has directed research on many advanced forms of presentation but, *Being Digital* is for those who have not yet discovered these advanced media. He states that the Internet was used by 20 million people in 1995 and, if this growth could be maintained, the number of users would exceed world population by 2003. In contrast to the slowness in exchange of ideas internationally in the early development of computing, the Internet is the most prominent phenomenon of the information society and allows the exchange of ideas irrespective of distance. It is just a network of computer networks which, thanks to the US military and academic support, has brought cheap, multimedia communications to many.

The data which is sent round the world can combine text, numbers, images, drawings, video and sound. It is the combination of these as multimedia and their transmission which is transforming business and will transform society. The break up of the Soviet bloc in the 1980s was partly due to communications by fax and satellite breaking the secrecy which had formerly been maintained. The truth of world events, such as the Chernobyl disaster, could no longer be kept from people by means of restrictions on travel and communication. Curiously, some of the first uses of communications were to enforce power, such as the electric telegraphs that were established along the railways of the British empire. Only ten years after Samuel Morse set up the first telegraph between Washington and Baltimore in 1844, one was built between Calcutta and Agra at the beginning of the unrest which led to the Indian Mutiny in 1857 [2].

The change of television to digital from its current analogue form, combined with the miniaturization of communications and flat screen displays, will result in some novel devices. Cable television will allow interaction so that viewers can join in games shows or vote in debates. Portable digital telephones will have greater intelligence and display screens so that they can access the Internet from anywhere. People may carry miniature digital video cameras with radio links for security. As recently as the early 1990s the fixing of locations using global positioning satellites involved large dish receivers and a personal computer. A mail order catalogue received in 1997 offered a hand-held global positioning device, locating any point on earth by reference to satellites, for less than £200. This is very new technology and of enormous significance for navigation, and was offered in the same catalogue as various ingenious, and often useless, devices such as shoes with spikes for aerating the lawn, or an appliance for massaging away a double chin.

2.4 *Communications causing fundamental change*

Rather than following individual pioneers or inventions leading to the information society, it is more productive, in anticipating the future, to consider broader trends. No one can predict who will contribute particular enhancements to technology but there are patterns which will persist in the future just as they have evolved up to the present time.

The conversion of data from analogue to digital form is one of these. Miniaturization of processing and storage is another. The most significant trend in communications is the provision of higher bandwidth which allows more data to travel faster along a single channel. Optical fibre provides the most dramatic growth in capacity and is already used for most trunk lines but it will take a long time to replace the copper cable which currently links each house or office. Cable television is delivered over faster lines and is competing with telecom providers for local communications. Negroponte [13] notes a trend towards switching broadcast communications from cable and wired communications to broadcast. If there were enough frequencies available and interference and privacy were not important, more internal communications in offices would be broadcast reducing the problem of wiring in buildings. Power will still have to be delivered by wire until portable, rechargeable batteries become more effective. Telecom cables can deliver sufficient current to maintain telephones through power cuts while optical fibre cannot unless sheathed in copper.

Another trend in the information society is towards the electronic persona. Personal lifetime telephone numbers are already being offered and Email addresses are becoming essential for most businesses, journalists and many individuals. The personal World Wide Web page is the electronic equivalent of the CV (*Figure 2.4*). Telephone banking is another growth area. In future many peoples' electronic persona will persist wherever they may happen to live or do business.

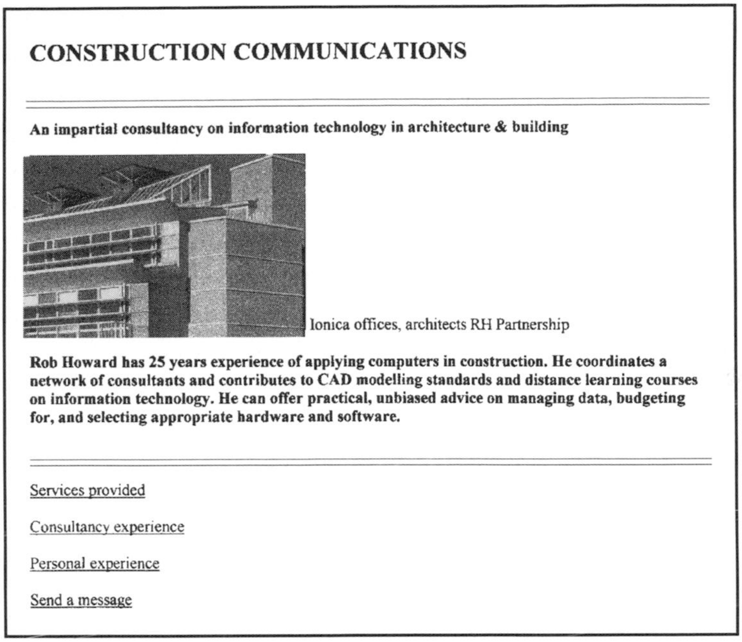

CONSTRUCTION COMMUNICATIONS

An impartial consultancy on information technology in architecture & building

Ionica offices, architects RH Partnership

Rob Howard has 25 years experience of applying computers in construction. He coordinates a network of consultants and contributes to CAD modelling standards and distance learning courses on information technology. He can offer practical, unbiased advice on managing data, budgeting for, and selecting appropriate hardware and software.

Services provided

Consultancy experience

Personal experience

Send a message

Figure 2.4 Home page on the World Wide Web. The Author.

Just as we are currently fixed in place by the need for letters to reach us or telephone calls to be taken at a number in a particular location, we are also fixed in time. 'I will call you at 8 o'clock', or 'Please sign for this registered letter', or 'The bank's business hours are 9.30 to 15.30' do not seem unreasonable at present. In future we will have more control over when and where we do things. Voice mail and the ability to transfer calls to different places, are already with us. Twenty-four hour telephone banking removes time restrictions on financial transactions. Television on demand will let us order the programmes we want when we want them. This time shifting may also help solve some of the problems of transportation with everyone currently wanting to travel at the same time.

2.5 *The take up of new technology*

The information society is the latest of a series of cultural and technical changes; from the agricultural revolution to the industrial revolution, and now the electronic revolution which has resulted in cheap, reliable hardware and software. The next stage is the emergence of the information society when data and communications, and their integration with the needs of people, are the main focus of development. However, revolutions do not change society overnight. Leading researchers and businesses may take up new developments rapidly and can establish their own vision of the future. Some visions may be seen by the rest of society as frightening, particularly if the intervening steps are not explained. It is these stages, of turning an idea into a commercial product, which are the most difficult to anticipate. IT has crept up on many reluctant users through their banks and insurance companies, and through intelligent devices in cars and washing machines. It will still take many years for conscious use of computers or the Internet to pervade all age groups and enter poorer societies. Access may not be the problem, with developments such as solar powered, radio phone boxes, foreseen by *Technology Foresight* [4], as an important development for the third world. There will always be those who do not use the technology available to them. Their wishes should be respected with their rights to information or communication capable of being satisfied by other means. Even with the fast communications of the future there will still need to be paper-based information, cash, voting booths and postal systems. Electronics cannot replace existing systems instantaneously.

Inventions supersede other inventions but take time to do so. The telephone eventually replaced the telegraph. Fax has replaced telex. The rate of these changes is much faster than those which went before, the replacement of canals by railways, for example. Technologically superseded systems, such as the canals, remain largely for leisure pursuits. How will society cope with all the information being directed at businesses and individuals, much of it unsolicited? Technology will be used to combat technology. Knowledge agents are items of software which will filter the electronic mail and other sources of data to accept only messages from known sources or to select information on specified topics. These will need to be set up carefully by each user to select the material required, add a little peripheral interest and exclude the unwanted. We can be sure that advertisers will have equally clever ways of getting round such knowledge agents. Soon agent will only speak unto agent.

The ability to use different time zones to process electronic information is starting to allow the sharing of work around the world. The stock exchanges of Tokyo, London and New York effectively do this with 24-hour trading overlapping the working hours in each of their time zones. Businesses in a hurry can do the same with projects worked on successively in different parts of the world to complete the task sooner. Similarly, work can be shifted to places where labour is cheap. The information revolution is therefore spreading resources around the globe and could help to even out some of the inequalities of wealth. The control of the multinational companies and the main software products is likely to remain in a few hands. These media moguls will become either the heroes or the villains of the information society.

2.6 The speed of IT evolution

No technology has been as fast in its evolution as IT. There are only 50 years between the 1940s, when it was estimated that the needs of the whole world would be satisfied by a handful of large computers, and the present time when small, reliable and cheap computers are essential to business and fast becoming a universal, personal accessory. Coping with this change has been difficult for many. The rate of obsolescence was so rapid that many used the excuse for not adopting IT that something better would arrive soon. It always does but, fortunately, there are plateaux

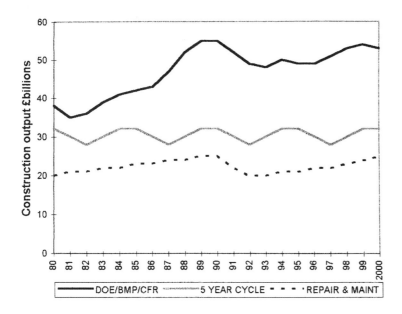

Figure 2.5 UK construction output 1980–2000. Department of the Environment, National Council of Building Materials Producers and Construction Forecasting & Research.

and certain technologies do stabilize for a period. It is at the beginning of such a relatively stable period when the technology should be adopted, but not before it has been proven, preferably by others.

For those in construction there is a particularly pronounced cycle of work load, which tends to have minor peaks about every five years and major ones every ten years (*Figure 2.5*). New systems are ideally introduced before the peak of each cycle, if resources allow, so that familiarity can be gained and training provided, ready to meet the demands of peak workload. IT cycles are much faster and move upwards in terms of performance for a given price. Most computer systems are written off over 2–3 years but, when they meet the business needs of a user, they should be adopted rapidly and provision should be made for replacing them later with what is likely to be even better technology.

3

Pioneers of computing in architecture, engineering and construction

History can be just nostalgia. It is not for this reason that the main developments of computer use in construction are summarized here, but to recognize some of the key contributors and the work of many others in evolving the systems that eventually started to increase the efficiency of building design and construction. History repeats itself and there are cyclical patterns which will recur in the future. Visions were presented many years ago which are still to be realized while other visions may have already proved false, leading to much work being wasted solving problems that were never seen as such by most architects and engineers. For all these reasons the history of these pioneering efforts is compressed into a few pages but only to direct this experience into more effective solutions for the future.

Before looking at the growth of IT serving construction, it may be useful to summarize developments in building design over the previous period and to speculate on how the masters of modern architecture and engineering might have used computers themselves.

3.1 *Design developments before computers*

Construction went through major changes in the period 1850 to 1950, particularly in the use of new materials and in attitudes to design styles. There was a series of new movements in architecture and engineering, symbolized by structures such as the Crystal Palace, which is generally attributed to Paxton alone, and the great French libraries of iron and steel, through Maillart's bridges of free form concrete, to the original thinkers of twentieth century architecture: Frank Lloyd Wright, Le Corbusier, Mies van der Rohe and Alvar Aalto. Would they have used computers if they were available? They certainly prepared the ground for a more dynamic approach to buildings.

> As it was for most architectural students of the 1960s, the great text of the period for me was Siegfried Giedion's *Space, Time and Architecture* [14] which shed light on the technical developments of the nineteenth century and inspired me to visit the great works of

modern architecture in France, Finland and the USA. Giedion's thesis on the perception of space over the ages was that the earliest buildings of Egypt and Greece were mainly seen from the outside and were symbolic and monumental. With medieval building, particularly in colder climates, came the increased importance of interior space and liturgical movement as in the Gothic cathedrals. The Renaissance put man at the centre of the universe and the newly discovered laws of perspective contributed to formal external spaces and composed groups of buildings.

The Victorian period was preoccupied with stylistic variations and it was the engineers and, even gardeners like Paxton, who are credited with the introduction of new materials, even if it has now been established that the Crystal Palace was largely a timber structure [15]. The twentieth century reintroduced the concept of movement through buildings and the interpenetration of internal and external spaces. How the designers of these buildings would have appreciated the dynamic presentation techniques of video and virtual reality!

New materials introduced new concepts of space through their innovative potential. Boulton and Watt's Salford mill of 1801 was a cast iron, prefabricated building of seven storeys, 42 × 13 metres (140 × 42 feet) in size. That illustrated is the later North Lane Mill in Lancaster, the refurbishment drawing for which won a CICA CAD Drawing Award for Geoff Leather in 1989 (*Figure 3.1*). Wrought iron enabled Labrouste to build the Bibliothèque Nationale in Paris with slatted floors so that light could penetrate several storeys of book stacks. Architects initially kept aloof from technological developments and it was men like Paxton, with his cast iron glasshouse system, and the contractor Fox, of the Crystal Palace of 1851, who showed originality in their prefabrication and design of the method of assembly on trolleys running along the iron and timber structure. He was up against a tight deadline and, rather than excavating to level the fall of several feet over the 1851 foot (564 metre) length of the building, he decided to erect the whole building on a slope with advantages for drainage and access at both ends for disabled people. A recent study of the extensive documents on this famous project by Professor Giovanni Brino [15] has revealed other features of this innovative project. There were telegraphic connections between the entrance doors, with the fire and police stations, and with the carriage park. It was possible to summon a carriage to the door on departure. The building was highly coloured and was burnt down twice after being moved from Hyde Park to Sydenham Hill.

The split between architecture and engineering goes right back to the formation of the Ecole Polytechnique in 1794 during the French Revolution and Napoleon's separately established Ecole des Beaux Arts in 1806. Structures were designed with increasing use of science and calculation during the nineteenth century and then decoration was added as a separate exercise. It was the experimental use of reinforced concrete by engineers such as Maillart in designing flowing curves for bridges, which brought the two disciplines closer. 'Calculation is the servant not the master' he is reported to have said [14]. It is doubtful whether calculation of these curved forms

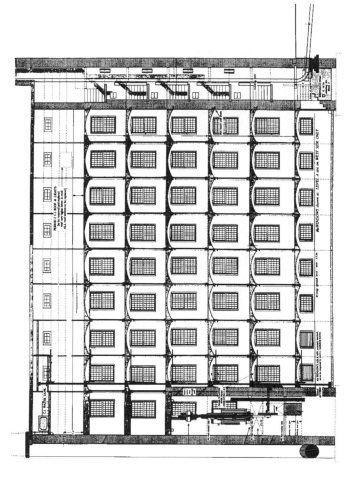

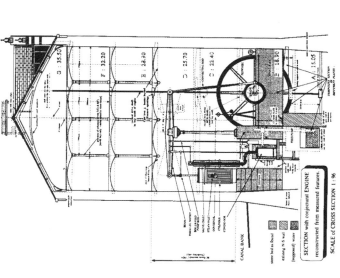

Figure 3.1 Nineteenth-century mill construction. North Lane Mill, Preston. Conversion drawings by Geoff Leather on Apple Macintosh winning the CICA CAD Drawing Award in 1989.

Figure 3.2 Millowners' Building, Ahmedabad, India. Ramp entering the building designed by Le Corbusier.

would have been possible at the beginning of this century. Even Felix Candela whose approach to experimental design of shell structures in the middle of this century was described by Sandy Wilson as 'Knock the shuttering away and run', could probably not design with great precision. Computers changed all this and, through sharing the data each needs, they also offer the opportunity to reintegrate the work of the engineer and the architect.

The arrival of Cubism and the enthusiasm of groups such as the Futurists for movement, influenced the architecture of Le Corbusier. Movement through buildings in an informal, non-axial manner and the exploration of space, particularly through interpenetrating volumes, became possible with new materials and a dynamic approach to buildings. Again, virtual reality techniques are only now making it possible to experience the richness of complex spaces before they are built. Le Corbusier's Carpenter Centre for the visual arts in Harvard has a ramp penetrating the building and joining adjacent streets.

In 1963 I was working in Boston and met Lionel March who was filming the Carpenter Centre as a way of expressing its dynamism. I later worked for B.V. Doshi, a former assistant to Le Corbusier, who had returned to practice in Ahmedabad in India where Corbusier's Millowners' Building also has a ramp which brings visitors into a louvred enclosure containing spaces divided by free flowing screens (*Figure 3.2*). Such movement through buildings was envisaged by designers but could then only be experienced after completion. New techniques were needed for presenting movement prior to construction.

Materials were a great influence on design until recently when buildings have become assemblies of many different parts, the material of which may not be finally determined until a contractor is appointed. Mies van der Rohe was the great master of steel construction. He carried on the tradition of skyscraper design in Chicago where his detailing of steelwork junctions and careful proportions link the small scale to the whole. He would have benefited greatly from computer draughting where a change at a detailed level, say of the size of a steel member, might affect the dimensions of the whole building. Another material, timber, became the hallmark of the Finnish architect, Alvar Aalto, who symbolized the emergence of Finland as a nation and expressed this through use of its native materials. His structures often use free form curves in two dimensions.

Figure 3.3 Advanced visualization. Saitama Stadium, Japan. By Hayes Davidson for the Richard Rogers Partnership, winner of the 1996 CICA CAD Drawing Award.

The Viipuri library, now in Russia, had an undulating timber ceiling when it was built before the second world war. In 1967 I went to Russia and passed through Viborg as Viipuri is now known. There were rumours that the library had been destroyed in the war but I was able to find the building, rather dilapidated, but still in use as a library. Another building by Aalto using free form curves in brick, is the MIT dormitory in Cambridge, Massachusetts. The advantage of the curved frontage along the Charles river is in the views along the river. These curves, which are also seen in Aalto's furniture and glassware, are symbolic of the Finnish landscape with its lakes and rocky outcrops. Setting out these curves can be quite a difficult problem for the contractor and, when visiting the offices of Reima Pietila, I was shown the drawings for the Dipoli building at Otaniemi. The large communal spaces have virtually no horizontal or vertical surfaces and the curved walls of granite boulders had to be dimensioned off an orthogonal grid, quite a complex process without the aid of computers.

3.2 *How computers were introduced*

Computers could have addressed a number of the needs of these pioneer designers but would they have helped their creativity? If these great men had used the technology available to us at the end of the century, might they have become obsessed with the computers themselves? I would prefer to think that, having the power to solve some of their problems, their creativity would have been even greater. The succeeding generation of architectural innovators in the UK represented by Norman Foster, Nicholas Grimshaw, Michael Hopkins and Richard Rogers, all adopted computers in their offices at the height of their careers, and this complemented their high tech style of repetition and clean lines. They present their competition winning schemes using advanced visualization techniques (*Figure 3.3*). These techniques, combined with the artistic ability needed, have produced a new specialist service of which Giuliano Zampi and Alan Davidson are leading exponents.

Presentation is only one aspect of Computer Aided Design use and the production of sets of drawings is the major application for CAD. Foster & Partners were originally advised on a CAD system for Stansted Airport by CICA. The Intergraph system was selected and used also to co-ordinate the services in the complex basement area. Keeping services in the basement and off the ceiling led to a very clear and uncluttered building. This was a management contract and the main contractors, Laing, were supplied with their computer systems for managing the project, an early example of a client wishing to ensure the use of new technology.

The masters of high tech are often said to have been brought up on bolted metal Meccano construction sets, while the post modernists of the 1980s learned design through the coloured plastic blocks of Lego. It is more likely that the colour capabilities which computers, gained during the 1980s, presented the opportunity for more colourful and decorative buildings. Certainly John Outram, in the Judge Institute of

Management at Cambridge University, used a multiplicity of colours and, where these are on curved surfaces, they were projected onto 2D drawings using CAD. The next wave of masters are likely to be hands-on users of even more advanced computer systems and will conceive their buildings with a continuum of information. Seamless transfer of all types of data between members of the project team will be assumed.

> In the USA the heritage of Mies van der Rohe has been carried on by Skidmore, Owings and Merrill. They are one of the few AEC firms to develop their own CAD system with integrated structural and services design software. I was present at their Chicago office for the launch of AESeries, based on IBM 6150 workstations, in 1989. This was at a time when their London office was growing and, when shown their several electrostatic plotters, I was interested to see drawings of Canary Wharf in London docklands being produced. I was told that they would then be sent to London by courier. Communications for transmitting large drawings electronically were not fully established at that time.

IT is concerned with manipulating similar aspects of building to those handled by architects and engineers over the ages. *Space* – distance is minimized by communications so that members of a project team do not have to be in the same location, and designs can be explained using advanced visualization. *Time* – through simulation of processes before they happen, dynamic forms of presentation and by sharing work world-wide.

> Bill Mitchell, in *City of Bits* [3], gives a wonderful example of space and time shifting when he describes being in the Europarc offices in Cambridge, UK, where the sun is setting over stone spires, and seeing, transmitted onto a computer screen from a camera in the PARC Palo Alto offices, the same sun rising over the dusty hills of California. The technology of transmitting images has taken a long time to evolve but is now dissolving distance and enabling organizations like PARC to establish greater unity between separate offices. It can also be used to reduce the separation between the site of a building and the offices of its designers.

3.3 *Application of the technology*

Human ability to imagine how a new technology might be deployed is amazing. Most of the applications of computers in construction were envisaged in the 1960s long before they could be delivered efficiently by the computers of the time. *The Architecture Machine* [1], published in 1970, chronicles the research carried out in the USA during the 1960s. Nicholas Negroponte dedicated it 'to the first machine that can appreciate the gesture'. Its arrival is still awaited.

Architecture in the 1960s moved towards rationalizing the design process, the universities having just freed themselves from the retrospective *beaux arts* approach to teaching. *Notes on the Synthesis of Form* [16] by Christopher Alexander looked at human patterns in the context of building. Sociological factors became more important. Alexander also wrote an article on 'the question of computers in design' as early as 1965. He also developed a computer program, HIDECS, with Marvin Mannheim which was concerned with the hierarchical decomposition of sets. This was applied to room association matrices to decompose them into groups of sets. This is an approach to automated design which was popular for tackling complex building briefs, and the automated bubble diagram for organizing the relationships between spaces is a facility still offered with some CAD systems and as part of the IAI's object modelling classes.

Other early publications on new technology include:

- *On the Evolution of Artificial Intelligence* IEE Symposium on Human Factors 1964.
- 'Computer-aided Design and Automated Drawings' *Architectural Record* October 1965.
- *Hologram Television* IEE Proceedings 1966.
- *Intelligent robots – slow Learners* Electronics 1967.
- 'Computerised Cost Estimating' *Architectural Record* March 1967.
- *Half Tone Perspective Drawings by Computer* AFIP Conference 1967.
- *Natural Speech for a Computer* ACM Conference 1968.
- *A Limited Speech Recognition System* AFIP Conference 1968.

There was already a *History of Computer Graphics* in 1969 and various papers on human/computer interaction, but the most surprising topic from this period was *A Head Mounted 3D Display* by Ivan Sutherland at the AFIP Conference in 1968. Virtual reality was conceived and demonstrated thirty years before computers were fast enough to exploit this vision effectively.

Another influential publication of the 1960s was *Architecture without Architects* [17] by Bernard Rudofsky and, although this was mainly concerned with structures created by animals or natural forces, it was symptomatic of a desire to learn about natural design processes and rationalize architecture. The term 'automated design' preceded 'computer aided design' since the early vision was to submit a design brief to a computer, with spatial requirements and relationships, and environmental constraints, and to generate bubble diagrams. These would then be turned into dimensioned plans, rationalized for construction, and would proceed to production drawings. This synthetic design process proved unrealistic since only a few factors could be quantified and these tended to dominate.

3.4 *University research in the USA in the 1960s*

The pioneering researchers in building design in the USA were at the Universities of Cornell, Harvard, MIT and Stanford. Use of the first graphical displays was explored with the help of Computervision, IBM and design offices such as SOM and Perry Dean & Stewart. Hardware was the limiting factor and the ergonomics of mice and light pens were tested as well as the use of slow storage tube displays driven by computers such as the IBM 1130. The first mouse used at MIT and Stanford was regarded as clumsy and cheap at $400. The graphics tablet was another input device introduced at this time and output was presented on the first pen plotters developed by Calcomp.

Sketchpad III by Timothy Johnson could, by 1963, display simple objects in three orthogonal, and one isometric, view on a single screen. The limitations were not on ideas of what to do, but on the volume of data that could be processed to represent buildings at all realistically. Another limitation was in disseminating the researchers' experience. Many papers were written and read enthusiastically around the world but the equipment to try out these early systems was confined to a few research laboratories. Using computers then required a knowledge of machine code or the primitive programming languages of the day. MIT was very foresighted in requiring all architectural students to take one term of computer programming from 1965 and in providing a time sharing computer to make this possible.

Automated design programs included GRASP (Generation of Random Access Site Plans) by Eric Teicholz at the Harvard Laboratory for Computer Graphics, and Comprograph, by the same team, working with Perry Dean & Stewart. These generated solutions to planning and building layout problems from a matrix of constraints. Design generation evolved into design appraisal in 1969 with RUMOR (random generation and evaluation of plans) which introduced the idea of generating many solutions and comparing their performance. This approach was taken up in the UK by Tom Maver at the University of Strathclyde when the ABACUS unit was established in 1969. Much of the early work in the UK and other parts of Europe was based on the American experience.

At the end of the 1960s, the URBAN 5 system represented the progress made in computer aided design. It was developed at MIT for exploring urban form and used a dedicated IBM 360/67 mainframe and an IBM 2250 graphics display with 8 000 bytes of local memory. Control of the display was by a purpose made set of function keys. Wire line perspectives were generated from automated assemblies of ten foot cubes representing elements of building. Its purpose was exploratory: 'to study the desirability and feasibility of conversing with a machine about an environmental design project'.

3.5 *Applying research in Britain*

These very early applications were purely for research purposes, to find out what the potential uses of computers might be and how they could complement human skills. The next decade developed this knowledge and related it to the need for productivity in the construction industry. Europe was still replacing buildings damaged in the war and coping with the needs of a growing population.

Structural engineering and quantity surveying were two areas of construction which identified potential uses of computers very early in their evolution. Complex structural calculations, particularly the analysis of finite element meshes, quickly became essential uses for programmable calculators and timesharing computer bureaux.

Ove Arup & Partners' experience of computing began out of necessity in the early 1960s, to analyse the shells of the Sydney Opera House with the help of Hewlett Packard calculators and university engineering departments. In 1964 they were the first consultants to install their own computer, an Elliott 803, for frame analysis. This was upgraded to an Elliott/ICL 4120 in 1966 which was also used for accounting. Input and storage were limited by a typewriter console and magnetic tapes. For the first time it was possible to work interactively on frames, highways and bridges, but through a specialist team, and the problem was integrating their work with the demands of projects for rapid results. The plotted deflection, shear force and bending moment, diagrams were of greatest use to the engineers. When in 1970 the 4120 was destroyed by fire, an IBM 1130 was acquired and a 3D frame detailing system was developed, but was not taken up widely by their engineers.

Their first CAD system tested was Applicon at £80 000 per seat at a time when an engineer cost only £2000 per year. It was rejected in favour of a more economical in-house development called CADRAW, first on the IBM 1130, then on a DEC 10 and finally on the PERQ graphics workstation. Then, in the mid 1980s CADRAW was phased out and Autocad adopted. David Taffs has led their IT department since 1966 and the practice has continued to be in the forefront of computing, even though it is still proving difficult to integrate applications with project demands.

In the USA the IBM 1130 became established in engineering consultancies and the CEPA (Computers in Engineering, Planning & Architecture) club was formed for these firms to pool their experience and software. Shortly afterwards the building services consultants, who used similar computers, formed APEC (Automated Procedures for Engineering Calculation). In spite of collaborating with both these organizations, I never did discover whether their acronyms being mirror images was coincidental.

In quantity surveying the obvious application was in generating a detailed schedule of items in a contract, the bill of quantities. This was an early database application and replaced a tedious form filling, summation and the printing process.

Quantity surveyors' use of computers was pioneered by Monk & Dunstone in the UK. A paper was presented to the Royal Institution of Chartered Surveyors (RICS) QS general meeting in 1961 by David Smart. They first used an Elliott 503 at a bureau and this read a magnetic tape with a complete bill of quantities library in under 5 minutes. Available

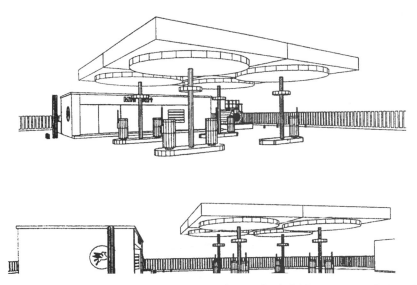

Figure 3.4 Mobil service station. Modelled in the GDS CAD system by D'Arcy Race with quantities derived by Monk & Dunstone.

storage was 8 Kbytes but the printer could produce 600 lines per minute, which was most impressive for the time. The main problem was in defining the 50 000 items in general use. The charge for processing and printing a small bill was about £75 and it was suggested that the RICS could rent such a computer and charge as little as £20 to its members.

Monk & Dunstone decided to buy their own Elliott 803 and, after two years' experience, K.W. Monk and P.H. Dunstone [18] assessed the contribution of the computer to the QS and included 'the re-employment, to the benefit of all members of the construction industry, of the information obtained during the preparation of the bills of quantities and from analysing costs'. This showed a perception of the value of the data they were processing for the whole project and yet we are still waiting for tendering against electronic data to become routine. In Germany, the StLb system of standard descriptions was introduced in the 1970s for government projects and this acknowledged the pioneering work of Monk & Dunstone in the UK.

A consortium of other quantity surveyors failed to mount a joint operation and MDA, as the firm became, set up a bureau service with Oldacres. They used various IBM systems, first a 360, then a 370 and eventually a 4300. A 1984 article by Geoffrey Ashworth [18] included later developments with microcomputers and linking bills with the GDS CAD system for Mobil service stations (*Figure 3.4*). Their systems were also used on the Lloyds building in the City of London with additional programs for measuring the services and assessing the cost implications of instructions for changes, rapidly.

A milestone in the generation of art by computer was the Institute of Contemporary Art's exhibition *Cybernetic Serendipity* [19] in the late 1960s. This included works by enthusiasts such as John Lansdown of Turner Lansdown Holt architects, and various mechanical devices which responded to sound or touch. There were drawings from another structural application by Ove Arup & Partners, of the façade of the ward block at

Northwick Park hospital designed in 1961. The apparently randomly placed structural mullions on this multi-storey building were calculated and placed according to the loads coming down from above. A rich pattern resulted and this expressed the indeterminate architecture of the designers, Llewelyn-Davies, Weeks & Partners. I completed my training as an architect by doing some of the staircase details on this project but computers had not yet been developed to carry out this chore.

The production of working drawings was not envisaged until the mid 1970s. Then the benefit was expected to come from re-use of tried and tested details. This benefit has since been limited by changing styles of building and the belief of architects that every project is different. In the early 1970s, there were major building programmes in Britain funded by public agencies who thought long term. Their ambitious targets could not be met without standardization and prefabrication, and the computer was seen as an aid to these processes.

3.6 *Computer systems for system building*

The construction industry of the 1970s was not concerned in the main with one-off projects, desperately seeking clients for these, or applying for lottery grants. Consultants did not have to submit fee bids or set up private finance for their projects. Building was largely devoted to meeting public needs for schools, housing and hospitals. Central and local government clients had budgets planned for years ahead and did not need to be wooed with photo-realistic perspectives or video animations. They wanted continuity and value for money and their design and construction teams were organized to provide this through prefabricated building systems and serial tendering. This offered the ideal environment in which to develop comprehensive computer systems.

An important event in bringing experience from around the world together with the early experience of pioneers in the UK, was an International Conference on Computers in Architecture at the University of York in September 1972 [20]. Contributors from abroad included: Terry Winograd and Nicholas Negroponte of MIT, John Gero of the University of Sydney, Bill Mitchell then at the University of California, Kaiman Lee of Perry Dean & Stewart, and others from Spain, France, Poland and Sweden. From the UK there were papers from research centres such as the Royal College of Art, University of Strathclyde, Bristol University, Portsmouth Polytechnic, Land Use and Built Form Studies at Cambridge University, Leeds Polytechnic, Loughborough University and the University of Edinburgh. Most significant at this formative stage were the outlines given of early building systems by their public clients: the Department of the Environment, West Sussex County Council, the Department of Health & Social Security and the Oxford Regional Health Authority, and by their consultants and the firms starting to provide them with computer services. This conference followed an exhibition at the University of Strathclyde in the previous year when some enthusiasts from the UK had demonstrated their software.

Pioneers of computing in construction in Britain were almost all at that exhibition and went on to take up the lead from the USA and develop serious applications to support the centralized building programmes, which were rare in the USA at that time. They were mainly based in universities where a few engineers and architects could see the potential of computers:

- Liverpool University was one where Arthur Britch led a research group into CAD which eventually resulted in both the CARBS system at Clwyd County Council and the RUCAPS system developed at GMW Architects.
- Leeds Polytechnic, under Bill Bradshaw, was working on 'the Model' which aimed to represent the whole building geometry and its performance.
- The Royal College of Art was led by Patrick Purcell, later to join the MIT Media Lab, who was collaborating with John Chalmers of the DOE Property Services Agency, on modelling government funded buildings.
- Edinburgh University had a Computer Aids to Architectural Design unit under Aart Bijl, which was starting work on housing layout and design for the Scottish Special Housing Association.
- Cambridge was represented by the new Design Office Consortium, formed by Applied Research of Cambridge, a commercial company started by the University Land Use & Built Form Centre, and the CAD Centre, a DTI research group which provided access to a timesharing Atlas II computer.

Architectural and engineering practitioners carrying out their own research into computers included: John Lansdown, Geoff Leather, Boyd Auger and David Campion of Cusdin Burden & Howitt, the last of whom did pioneering work on scheduling hospital fixtures and fittings before developing his own CAD system.

Boyd Auger is an architect and engineer who foresaw the potential of computers in design at an early stage. His book, *The Architect and the Computer* was published in 1972 [21] and describes work done in the late 1960s at Leicester University with Geoffrey Butlin, on the BAID system to aid the design of housing estates. Afterwards, at Imperial College in London, he helped Colin Besant and his team, who were developing the hardware for a computer draughting system, by flow charting the drawing process and helping to prepare the BASYS software required for use by architects for CAD. This system used a digitiser-plotter developed by the team for D-Mac Ltd, with a DEC PDP8 computer, and made the first use of a Tektronix storage tube display (*Figure 3.5*). He could see the possibilities for integrated software around the building systems of the time and the generation of quantities. He urged architects to explore the new technology without expecting much financial return and had no illusions about the proper role of the computer. 'An architect who possesses such a capacity should feel confident as a technologist and would thus be freed mentally to concentrate on his responsibilities as an artist'.

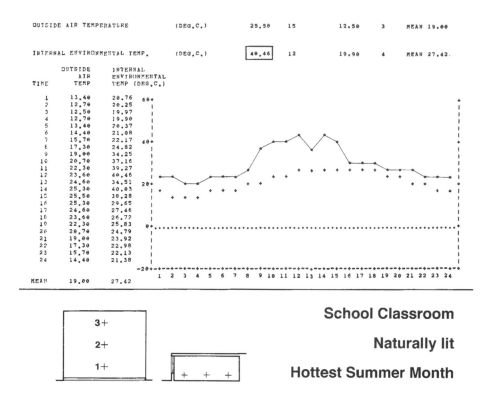

OUTSIDE AIR TEMPERATURE	(DEG.C.)	25.50	15	12.50	3	MEAN 19.00
INTERNAL ENVIRONMENTAL TEMP.	(DEG.C.)	40.46	12	19.90	4	MEAN 27.42.

TIME	OUTSIDE AIR TEMP	INTERNAL ENVIRONMENTAL TEMP (DEG.C.)
1	13.40	20.76
2	12.70	20.25
3	12.50	19.97
4	12.70	19.90
5	13.40	20.37
6	14.40	21.08
7	15.70	22.17
8	17.30	24.82
9	19.00	34.25
10	20.70	37.16
11	22.30	39.27
12	23.60	40.46
13	24.60	34.51
14	25.30	40.03
15	25.50	38.28
16	25.30	29.65
17	24.60	27.46
18	23.60	26.72
19	22.30	25.83
20	20.70	24.79
21	19.00	23.92
22	17.30	22.98
23	15.70	22.13
24	14.40	21.38
MEAN	19.00	27.42

School Classroom

Naturally lit

Hottest Summer Month

Figure 3.5 Output from the Cambridge Environmental Package, 1974. Internal environmental and outside air temperatures in a school.

3.7 *Applications for contractors and surveyors*

Contractors' main applications were in accounting and involved processing data from many contracts and consolidating the results. This was typically carried out by batch processing but the speed of receiving the consolidated accounts was important for their businesses. Most employed bureaux but Costain and Mowlem shared an ICL 1900 Series computer and offered time to others via their Computel service. Local authorities usually had computers, mainly for the benefit of their Treasurer's Department, and early attempts to use these for building design or analysis were given low priority. They also had a commitment to the building systems that were established by consortia to meet the need for schools: CLASP, SCOLA and SEAC were three of these and the development of bills of quantities programs by the Local Authorities Management and Computer Committee (LAMSAC) was largely based on counting the limited range of components in the CLASP system. Derbyshire County Council quantity surveyors under Vernon Foster went on to produce working drawings as a by-product of this application. Private housebuilders were not so forward looking at this time but the Scottish Special Housing Association used a concrete building system which lent itself to computer aided design and scheduling. System building declined following such prominent disasters as the collapse of the Ronan Point tower block and the rapid spread of fire in a CLASP built student hostel in Paris.

Energy saving became an important consideration after the 1973 rise in oil prices. Heat loss calculations were among the analysis programs being attached to design systems since environmental effects are linked to building shape. The Land Use and Built Form Studies centre developed the Cambridge Environmental Model which linked heat loss to summertime temperatures, daylight, artificial light and even sound penetration. Dean Hawkes, who designed the model, always maintained its purpose was to aid designers to learn how to minimize energy usage and services systems while providing comfort conditions. This concern is borne out by current thinking on energy systems, which use fluid flow dynamics to simulate natural stack ventilation, and place more emphasis on control systems. Parallel calculations were carried out to determine the need for air conditioning in summer and the trade off between natural and artificial light, where the gains from light fittings were fed into the heating calculations (see *Figure 3.5*). At that time there was no modelling of plant and services so the environmental model could be used at an early design stage and the building could be modified to improve its performance. Soon building services design software was being written at Oscar Faber & Partners and by Tony Baxter at Sheffield Polytechnic. More complete environmental models resulted from combining all these applications but integration was then only feasible between a limited range of functions.

3.8 The first users of CAD systems

The UK has always been fortunate in sharing a language with the USA and in being the first landing point for new developments from across the Atlantic. When the pioneering work of the American universities became commercial, it was first exported to Britain before being launched into the rest of Europe. John Laing, for instance, was one of the first European users of a dedicated CAD system, Computervision, in the mid-1970s. The Percy Thomas Partnership invested in the ARK 2 system from Perry Dean & Stewart. Both these firms gained exposure and experience from these expensive and limited systems which they have since turned to their advantage. Through the 1970s other European countries lagged behind Britain although they were quick to learn from its experience during the 1980s.

The UK Department of Health & Social Security (DHSS) was charged to produce 300 district general hospitals by the end of the century. It planned a very systematic approach and had collected much data on movement in hospitals, standard libraries of fixtures and fittings, and department plans. Such large and functional buildings needed systems. My own interest in computers resulted from working on hospital laboratories, such as the Institute of Neurology in London, where the scheduling of furniture and equipment, selected by clients from a mock-up of a laboratory bay, was a repetitive chore. This seemed an ideal application for computers.

The DHSS building system to meet this long term need was called Harness and was comprehensively planned on a 15.3 m square frame of precast concrete elements with courtyards to provide natural ventilation. Standard departments and room layouts were planned within this. The geometry was simple to model and design aids developed by Applied Research of Cambridge (ARC) and others included circulation and lift analysis, sunlight and daylight, heating and cooling, site analysis and simple perspectives. The systems were developed at the CAD Centre using its timesharing service but output, other than that typed on a teletype console or storage tube display, linked by a slow, dial-up telephone line, had to be collected from the computer centre. Data preparation and input were primitive with little automatic checking so that the plotted drawings, each of which took up to one hour, would be the first opportunity to spot flaws. Harness was an ambitious project carried out by a large team over several years and put through exercises with hospital designers. It suffered from rivalry between the Department of Health and the Regional Health Authorities who were responsible for individual projects. Inevitably the political will to carry it through eventually declined and, apart from a few conference papers, the value of this work was only realized through its developers going on to produce more generalized systems.

While these major building systems were being developed, individual applications programs were becoming available through on-line bureaux to introduce a larger number of firms to the possibilities of computing. Surprisingly the list of types of application recorded in a survey by ARC and Systems Programmers Ltd for the Department of Trade & Industry in 1973 [22] was little different from lists maintained by the Construction Industry Computing Association in the 1990s.

Software categories

DTI survey 1973	*CICA Software Directory 1996*
In-house management	Office management
Brief analysis	
Schematic design	Computer-aided design
Design aids	
Production information	Draughting systems
Sites and land use	Sites, mapping and GIS
Detail design	Structural analysis and design
Performance evaluation	Environmental and energy
Detail design	Building services
Costing	Costs and quantities
	Construction management
Project control	Project management
Buying/delivery	Manufacture and supply

There was rather more emphasis on design and environment and fewer management programs. Word processing and spreadsheets did not exist in their present form but this list of applications is evidence of the ability of users to anticipate how a new technology would be applied. It was then a matter of producing efficient software which would exploit the reducing cost of hardware and could be applied to buildings of all types rather than just those designed within highly rationalized systems.

Constraints on the use of computers in the 1970s were not due to lack of imagination, but to hardware which was slowly evolving towards separate, reliable and affordable minicomputers and then, in the early 1980s, cheap personal computers. Software was written from scratch, often by those who had learned computing doing research in a variety of disciplines, using machine code or Fortran, and tested by trial and error.

> The Harness project was directed at ARC by Ed Hoskins who was qualified as both a dentist and an architect, and had four project leaders: Keith Wylie was a computer scientist who went on to become a barrister, Paul Richens a natural scientist and architect who became director of the Martin Centre in the Cambridge University Department of Architecture, Ian Steen who returned to architectural practice and became known for designing doctors' surgeries, and Peter Hayward, a geneticist who went on to make specialist hardware. Although their work never got the publicity it deserved, it was developed into a more widely usable system for the Oxford Regional Health Authority and eventually into the General Draughting System which had much success during the 1980s.

OXSYS was the brainchild of Malcolm Jones who went from the East Anglian to the Oxford Regional Health Authority where a prefabricated method of construction, known as Oxford Method and based on steel frame and panels, was in use. It was not such a large scale system as that designed for Harness, which assembled pre-planned departments, and stood a better chance of public acceptance. It could handle any size of project and was more flexible within the constraints of the orthogonal geometry to which many systems were limited at that time (*Figure 3.7*). Even so the possible combinations of four partitions of two different thicknesses coming together centred on, or face on, to crossing grid lines, created a very large number of possibilities, each of which would have to be detailed in a comprehensive model. Components which were not rectangular were modelled within rectangular boxes, as if their plan and elevations were pasted onto the faces, a clever technique which generally worked in plan and elevation but was rather less convincing in section or perspective.

3.9 New hardware and its social effects

These health system developments, together with those for housing and schools, laid a foundation for future systems but are often forgotten today when there is talk of comprehensive building models or integration of applications. Long-term building

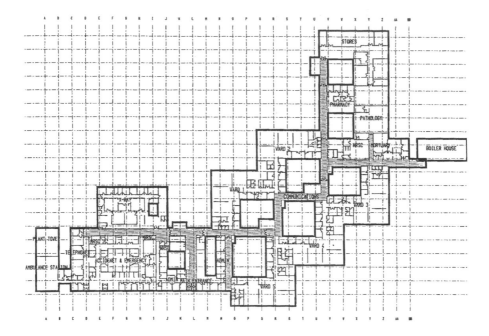

Figure 3.7 Final plan of the Milton Keynes District General hospital which was designed with the OXSYS system. Oxford Regional Health Authority, drawing supplied by Steve Race.

programmes declined and system building became less acceptable to the public. The interest of construction in computing had been stimulated and all that was needed to exploit this was a range of general purpose programs for one-off buildings and more accessible hardware. The attention of those who had pioneered system building applications turned to CAD software which could draw anything to be found on a drawing board.

Time sharing computers, paid for according to the resources and time used, with response dropping at busy times, were never the ideal means of developing software. Minicomputers from DEC and Prime arrived in the middle of the 1970s and were followed by networked graphics workstations. Software houses and universities could purchase these but users of their software in industry had to have access to similar machines. Standard operating systems, such as UNIX and CP/M, prevailed at the end of the decade and software became more interchangeable.

There was a growing number of desktop machines before the personal computer arrived, and these were typically used for engineering calculations and came from such companies as Hewlett Packard and Olivetti. For training and home computer enthusiasts, the Commodore PET and the Apple II came at the end of the 1970s. Contractors were still having accounts and payroll processed by bureaux as a batch operation but this could take several days. For effective office and project management, an interactive system, preferably a dedicated computer, was essential to speed up the generation of valuable management data.

As computing emerged from the control of a few specialists, training became a problem. University departments in the UK were crying out for more equipment and hands-on access was still confined to a few researchers. In 1980 the Design Office Consortium became the Construction Industry Computing Association following a government report recommending it as the national centre for computing in construction [23]. It had established a reputation for good, impartial advice by evaluating applications software for the DOE Property Service Agency and publishing comparative reports on various types of application.

CICA was charged with providing training and, over the period 1981–4 in conjunction with the Builder Group, it provided hundreds of architects, contractors, engineers and surveyors with hands-on access to Apple II computers over a three-day period. These courses were devised by Simon Bensasson, who later became head of research at the European Commission DG XIII, and provided a theoretical introduction to microcomputers and simple programming exercises in Pascal. Few delegates intended to become programmers but the exercises gave an insight into how computers worked. The second half of each course was dedicated to applications appropriate to the professions present. Those demonstrating their software included: John Lansdown on CAD and, later, expert systems, Geoffrey Hutton on information systems, Tony Baxter on building services and Peter Burberry on environmental analysis. At the end of each course delegates could hire the computers to consolidate what they had learned and demonstrate this to their colleagues. Many of those who went through these microcomputer workshops went on to become IT specialists in their firms or to develop software themselves.

The accelerating progress of computing in construction during the 1980s is more fully documented in institutional surveys and conference papers such as those from the series run annually by CICA and RIBA Services on CAD at the RIBA. These started to look at the aesthetic and social effects of computers but there had been rather more concern about effects on employment in the 1960s.

Geoffrey Hutton was a pioneer in the use of computers for photocomposition in the mid 1960s. This threatened the powerful printing industry of the day and, when his production of product manuals for materials suppliers such as Cape started to include drawings, they also threatened the role of the drawing office. He, and his partner, Michael Rostron met as research fellows at the Architects Journal in 1961 and set up a programme of 'elemental design guides', a knowledge-based series arranged on the new SfB classification system. Geoffrey's long experience of databases and preparing publications for building materials producers and information services, has taught him that classification systems are less important, in machine-based systems, than indexing and coding. This resulted in the development of Keycodes, a method of encyphering concepts in the material for retrieval. This is a multilingual, interlexical tool forming part of the reference engine in Hutton + Rostron's information systems. His work in the 1960s on electronic publishing from databases has now grown into the Alpha DIDO on-line business support system, developed with the aid of European Union funding. This is a fully transactional teleworking system hosted at his offices in Surrey on a Sun SPARC station supported by a Vax computer. His other computers include early machines such as PDP 8 and 11s from the 1970s, kept for historic interest, while a network serves his other businesses. These include the Curator on-line monitoring system for environmental investigations.

Hutton + Rostron's first computer-driven photocomposition was for Cape Asbestos in 1966. This used punched paper tape and could produce schedules of cladding components, assembly instructions and digitized drawings. Unfortunately the print unions were against this technology and insisted on resetting the first Cape Technical Manual by hand. In the early 1970s computer graphics became feasible using a Ferranti system printing onto 35 mm aperture cards, but the drawings were seen as a threat by the client's drawing office staff and the project to produce job drawings on demand was abandoned. By the time computer graphics had become more effective the inevitability of computers was generally accepted and the need for business to compete internationally prevailed.

The 1983 RIBA CAD conference was on 'How CAD will affect your office'. It was chaired by Bryan Jefferson, a past president of the RIBA, and was concerned with the social effects of computing. This was at a time when computers were creating more jobs than they removed and the effects noted were: the need for good management and training, by the Building Design Partnership, a salaried architect's view from the Greater London Council, who had a large IBM CADAM system, on the difficulty that older people were having in adapting to the new technology and, also presented, were the results of a study carried out by CIAD in the Netherlands. Few of these effects were insurmountable. David Taffs found the main difficulty in integrating computers was in offering a fast service from a specialist computer team to engineers needing to respond quickly on projects. Mike Stace, of Dudley Marsh architects, stressed the opportunities provided for building models in the computer before construction started, reducing errors and being competitive.

Development soon concentrated on general applications for the IBM Personal Computers, which arrived in the UK in 1983, and their growing ability to handle draughting systems. The evolution of these is covered in Chapters 5 and 6. In the 1990s there is a new generation of researchers and a tendency to ignore the experience of the pioneers. Some of the early visions for computers have been reinvented and this history may have some useful lessons for those who were not involved then. **The fact that some of the early systems did not achieve their ambitious goals, often as a result of the fluctuating cycles in construction, should not deter future attempts to attain the goals of integration and common building models with the vastly improved IT facilities now available.**

4

Changing conditions in building design and procurement

The prevailing view of the construction industry is that it changes slowly while other industries are evolving rapidly. The IT industry changes faster than most. It barely existed until the 1970s and, by 1994, the level of consumption of IT products in the UK had reached 4% of GDP. Earnings from IT were lower at 3% but were expected to grow to 10% by 2005 [4] which would make it the largest industry unless construction returns to its previous high level of 10% of GDP.

4.1 Some elements of change between 1970 and 1995

Construction in Britain has been forced to change by increasing its productivity and exploring new ways of procuring buildings. Until recently little of this change has been caused by IT but rather by economic factors and cultural attitudes towards the design of buildings. These changes have been significant and it is illuminating to compare some aspects of construction over the period 1970 to 1995. The former date represents the advent of timesharing computers offering access to the first users in construction. The latter may herald another wave of change in construction resulting from the Latham Report [24] and the passing of the Housing Grants, Construction and Regeneration Act which followed from it.

Workload

At 1990 constant prices, construction output in the UK was £38 billion in 1975 and £50 billion in 1995. Between these dates it fluctuated on a cycle of about five years and the low and high points were £35 billion in 1981 and £50 billion in 1990 (see Figure 2.5). Coping with these fluctuations had an effect much greater than that caused by computers. IT has started to contribute towards the 30% savings called for in the Latham Report but, what has been saved, has been more a result of harsh trading conditions than of a planned use of technology to add value.

The balance of work between the public and private sectors and between new work and repair and maintenance changed significantly during this period. The following percentages are from the Department of the Environment reported in *Building Britain 2001* [25] and from Construction Forecasting & Research reported in *Building* [26] (*Table 4.1*).

Table 4.1 Percentage of UK construction workload in different sectors, 1986/1996

	1986	1996
Public new work	24	15.7
Private new work	42	38.3
Repair and maintenance	33	46

The increased value of existing property and the higher standards of finish expected by occupants have resulted in a higher level of spend on refurbishment. Only if the Private Finance Initiative fails will there be a large scale reversal of a diminishing public sector and a return to commissioning buildings directly.

International competition

With IT making industries like construction more global it is important to compare the cost of building in different countries and the size of their construction industries (*Table 4.2*). The indicator used in the W. S. Atkins report for the European Union *Strategies for the European Construction Sector – a Programme for Change* [27] to compare costs of building in 1990 minimizing the effects of exchange rates, was Purchasing Power Parity (PPP).

Table 4.2 Comparative costs of building related to UK costs, 1990

Country	Market rate (%)	PPP (%)
UK	100	100
Germany	97	96
France	81	101
EC median	82	98
Japan	86	77
USA	73	108

So the UK was then more expensive than the median rate for the European Union but significantly cheaper than the United States, in spite of reports comparing costs on comparable projects and the greater speed of building in the USA. Japan is the cheapest of these in which to build but this probably ignores land values.

The approximate national proportions of the £500 billion European construction market show greater variation than the relative sizes of national economies (*Table 4.3*). These figures are from an RIBA leaflet 'Impact of the single market' and *Technology Foresight – Construction* [4].

Table 4.3 Percentage of the European construction market, 1985/1993

Country	1985	1993
UK	16	10
Germany	30	30
France	18	16
Italy	17	16

In spite of German reunification its share has remained the same while that of the UK has diminished even allowing for the additional members of the European Union.

Contracts

In 1970 most contracts were competitively tendered against Joint Contracts Tribunal (JCT) forms of contract based on bills of quantities. There was also serial tendering to reflect the longer term public building programmes. Since then new forms of procurement have appeared alongside JCT: construction management, management contracting, and design and build. The comparative advantages of each of these were quoted in the Latham Report from CUP Guidance Note 36 [28] (*Table 4.4*). Additional criteria for the advantages of each for use of IT might include the stage at which the contractor was appointed and responsiblity for project and data management.

Table 4.4 Comparative advantages of different forms of contract, CUP 1992

Objective	Traditional	Construction management	Management contracting	Design and build
Early completion		*	*	*
Price certainty	*			*
Quality of design	*	*	*	
Avoid costly change	*	*	*	
Complex buildings		*	*	
Single contractual link				*
Design team to report	*	*	*	
Transfer complete risk				*
Cost recovery	*		*	*
Contractor input to design		*	*	

The latest form of contract, endorsed by Latham and designed for modification by computer, is the New Construction Contract originally devised by Dr Martin Barnes based on the experience of simpler forms of civil engineering contracts.

Fee scales

For professional practices there were minimum fee scales in the 1970s and these seem generous in retrospect. The UK government of the 1980s was determined to destroy monopolies, particularly the one they believed was maintained by architects, and they banned minimum fee scales. Ruthless competition set in and had few ill effects until the boom of the late 1980s collapsed when the quality of much design work declined. CAD began to be seen as essential to working within lower fees or submitting the growing amount of speculative work more efficiently. By the end of the 1980s CAD was also becoming essential to produce sets of construction drawings and incorporate changes in design without incurring large extra costs. In the 1990s consultants spend much of their time bidding for projects and assembling finance. Contractors also had to survive on low profit margins and economies were found through keener purchasing, delays in paying subcontractors and the use of IT systems.

Specialist consultants

The construction professions had been fairly stable from the establishment of the engineering and surveying institutions in the nineteenth century up to the 1970s. During this period building had increased in complexity and, after 1970, more specialist groups emerged, particularly specialist subcontractors, project managers and facilities managers. Building began to be seen as dependent upon a flow of information right from the briefing stage to the operation of the resulting building. The need for good communications to keep the client informed of progress and to pass project information from the growing number of specialist designers to contractors and their many subcontractors, and on to building occupants, has become more essential as complexity has increased. The result is larger project teams with greater problems of co-ordination and more information for each to handle. The solution may lie in the appointment of information, or document, managers who maintain a central electronic resource accessible to all. The need for sending copies of everything to all parties becomes unnecessary when the latest data can be retrieved at any time. This is feasible for new projects planned on this basis, but electronic records of existing building stock can take many years to create.

Building types

The most noticeable change in the types of new building being designed is the switch from public clients and social provision to the private sector and new commercial and industrial developments (see Table 4.1). About half of the £50 billion construction workload is now spent on repair and maintenance. In 1970 the population of the UK was growing more rapidly and replacement of old buildings was still a priority. Since then it has been left to market forces and these have not always met the need. It is likely that, before the end of the 1990s, the use of private finance to support vital services will need to be reassessed and the building industry could be asked to gear up its output rapidly.

With the lack of training provided during the recession there may need to be a return to the system building programmes last seen in the 1970s, with prefabrication on a more human scale to meet the diverse needs of customers. Funding from the National Lottery seems set to continue and will provide a welcome stimulus to cultural and charitable projects. However, the supply of affordable housing and up-to-date health and education buildings needs a new public commitment before the end of the century.

Types of client
Another change in procurement is in the nature of those commissioning buildings. Privately commissioned houses were still quite common in the 1970s. Architects were more involved in mass housing, usually for public authorities. Now this market is dominated by mass housebuilders and their marketing specialists who design to meet conservative preferences. Housing Associations have carried out most of the social housing but their grants have been reduced and the quality of design has suffered. Of the former government clients, only the Ministry of Defence is still a leading procurer of buildings. New offices, even if let to public authorities, are developed by the private sector.

Designers and contractors have often had to become developers themselves to secure their flow of work. Some industrial clients, such as the major retailers, utilities and transportation firms, have become expert clients with detailed briefing and specification manuals for their consultants, and they are likely to promote change in the way their buildings are procured and produced. They see construction as the way to expand their market share and the building process becomes a part of their own business process. *Building* [29], which provides excellent reports on the business of construction, publishes a monthly list of leading client organizations. In the year to September 1996, the first three were retail chains followed by the Ministry of Defence, BAA and Tarmac. A similar list from 1970 would have featured largely central and local government bodies.

Documentation of buildings
In 1970 the types of document required to build in traditional construction were well understood and drawings, bills and specifications could be read together. They were organized by different classification systems but used the traditional trades and terms that were familiar to all. With increased variety in forms of construction and different types of contract, there was a need for greater co-ordination of documentation.

Co-ordinated Project Information (CPI) was a joint project of the Association of Consulting Engineers, the Building Employers Confederation, the Royal Institute of British Architects and the Royal Institution of Chartered Surveyors. It produced codes for drawings and specifications, and a new Standard Method of Measurement, all classified by a work group based classification system called Common Arrangement. This provided the linkage between related information in each document through well researched sources of data such as the National Building Specification.

A report from the University of Bath [30] assessed the level of adoption of CPI following the work done there on communications between members of the project team. They found that a high proportion of consultants claimed to use CPI but that they were probably employing their own form of co-ordination rather than using the CPI structure to the letter. An awareness was created of the possibilities for smoothing the flow of information and this is starting to be applied through recent IT developments such as document management. As a result of the increase in efficiency the industry has had to deliver during the 1990s, the flow of information, control of revisions and the team members to whom it is issued, have to be recorded accurately. Chapter 8 deals with the facilities offered by IT for achieving the 'less paper' office.

The building process
Mechanization has transformed the building site from a place where much manual labour, formerly employed directly by the main contractor, has been replaced by subcontracting and a greater dependence on high technology plant even if there is not much use of autonomous robots yet. There is a constant search for new materials and fashion often dictates which of these are successful and can compete with tried and tested products. Construction management is now a highly sophisticated process with information on progress from each site, and its profitability, transmitted to the contractor's head office daily and consolidated to show the board whether they are meeting their profit targets.

Partnering is proposed in the Latham report so that greater trust is developed and profits do not have to be recovered in claims or litigation. When profit margins become more generous again, this will be tried out and, alongside partnering of business processes, there should also be sharing of project data. In the 1990s we have the technology for the whole team to access common project data and, with greater trust and appropriate controls, this could help reduce the 25% of site problems which the Building Research Establishment reported in 1981 [31] were caused by poor information.

These are just some of the fundamental changes which the construction industry faced between 1970 and 1995. They have provided opportunities for the faster emerging IT industry. The rate of such change has already been hastened by the IT systems which have been widely adopted. This rate of change is bound to accelerate as a result of more radical alterations which will affect not just the isolated tasks which lend themselves to automation, but the whole flow of information throughout a building project.

4.2 *Opportunities for information technology*

It is often difficult to separate cause and effect. For example employment in the UK construction industry fell from 1.8 million in 1989 to 1.4 million in 1995. Did IT contribute to this or has it only allowed a similar volume of work to be produced by fewer people? Until the 1980s computers had created more jobs than they had removed. They then started to change administrative organizations, such as banks and insurance companies, radically. In construction, jobs were still being created, often by building professionals setting up software houses or training as IT specialists. The UK IT industry was estimated to employ 365 000 in manufacturing, 130 000 in software and services and 426 000 in IT vendors and users, but there is some overlap between these figures [4]. As well as securing more equipment for undergraduate courses, a number of MSc courses were launched in universities largely for mid-career students seconded by far-sighted employers.

At the end of the 1980s building boom, computer systems for draughting and accounting became sufficiently effective to reduce the number of people needed to carry out these functions. Word processors had reduced the number of secretarial staff beforehand, but these more complex applications took many years to threaten design and management jobs. Nigel Cross of the Open University wrote *The Automated Architect* [32] in 1977 and quoted a number of promises and threats resulting from use of CAD:

- Real time walkthroughs – greater stress on the designer.
- Much time would be needed for maintenance and repair of computers.
- Working from home would be attractive – staff reductions needed to pay for computers.
- Improved design through analysis – quantitative elements outweigh the qualitative.
- Deprofessionalizing design – greater involvement of lay people.
- There was little evidence of different characteristics in buildings designed with CAD.

Several of these concerns have not proved well founded, largely through unforeseen cost reductions and increased reliability of computers. The take-over of design by lay people may yet happen through greater use of personal computers and virtual reality. The general conclusions of *The Automated Architect* were that, following lengthy analysis of design problems tackled by people and by software such as SPACES from Abacus, there were 'no particular advantages for the use of computer aids to solving a building design problem'.

Figure 4.1 Gathering of international CAD experts by Loch Lomond in 1990 to celebrate 21 years from the founding of Abacus at the University of Strathclyde. From the left: student and Gianfranco Carrara, Ivan Petrovitch, Bill Mitchell, John Gero, Tom Maver, John Lansdown and Wojciech Gasparski.

Nigel Cross reviewed these predictions at a gathering of some of the pioneers of computing in building design to celebrate 21 years since the founding of the Abacus unit at Strathclyde University, in 1990 (*Figure 4.1*). While his conclusions were obvious to the twelve people present, other, new predictions were made. There is still a wish to define the design process better so that it can be simulated by computer. John Gero, of the University of Sydney, was applying neural networks to cognitive models of designers. Bill Mitchell, then at Harvard University, predicted: design oriented geometric modellers, feedback from analysis while designing, control strategies using parallel processors to co-ordinate design work carried out simultaneously on the same project. David Taffs of Ove Arup & Partners was more sceptical of the benefits computers had brought but admitted that his business could not manage without them.

4.3 Changes in employment and buildings

With most types of business, particularly those involving routine administrative work, reducing staff to be competitive through increased automation, so construction reduced its staff to the number it could afford to employ in the 1990s, and so had to make greater use of computers. The first professional offices with one computer per staff member appeared although *Building on IT for Quality* [33] showed that, in 1993, the median provision in larger architects, engineers and surveyors was one computer for every three members of staff and, for contractors, it was one for every five.

Even partners and directors had to learn to use the personal computers with which they were provided. Construction industry jobs are rarely just administrative, so the staff who remained had to be good at bringing in the work, designing or managing, and increasing their efficiency through use of the systems at their disposal. Many technicians, who had been the most productive because they used CAD regularly, left the architectural and engineering offices which had employed them at this time, and qualified architects and engineers had to learn to use the CAD systems in which they had invested. This resulted in eliminating the separation of design and draughting and led to a more strategic use of design software. CAD then started to become an aid to the whole design process rather than just a means of producing drawings.

I organized a series of conferences on aspects of CAD at the RIBA through the 1980s and, in 1982, the theme was 'Buildings designed with computers'. What effect was CAD having on the appearance of buildings? The *Architects Journal* wrote this up in January 1983 [34] and concluded that, at that time, it was affecting the process of design rather than the resulting buildings. The contributors to the conference included: Roger Cockle of the Percy Thomas Partnership with their use of the ARK2 system on Leicester Hospital, David Campion of Cusdin Burden & Howitt on a terrace housing development, John Gibbons on the Milton Keynes Hospital designed with OXSYS, and Colin Connelly of BBCP on housing rehabilitation using RUCAPS. Their general conclusions were that CAD was not being fully exploited yet and therefore had had little effect on the design of these buildings. However the Milton Keynes hospital had gone through more design options than normal and this was largely possible owing to the facilities of OXSYS for comparing performance and cost, and redrawing (see *Figure 4.2*).

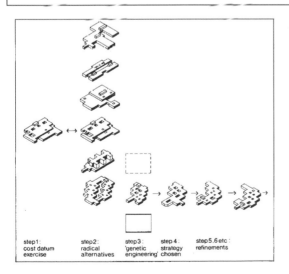

step 1:
cost datum
exercise

step 2:
radical
alternatives

step 3:
'genetic
engineering'

step 4:
strategy
chosen

step 5,6 etc :
refinements

Figure 4.2 Series of plans showing the design developments of Milton Keynes District General Hospital.

Since then the possibilities for design have widened with the ability of CAD systems to draw tent structures, polyhedral buildings and those with non-circular curves. Two examples illustrate this: Lancashire County Council was experimenting with polyhedral school classrooms made from fibreglass panels. They had problems in drawing the complex elevations and sections, and perspectives were virtually impossible. One afternoon at the CAD Centre in Cambridge, using their THINGS and HIDDEN LINES software, resulted in a series of views produced by CAD with relative ease. Unfortunately the oil price rises which followed made this form of construction uneconomic. Another example, quoted by John Hare, then at Arup Associates, was a building for British Sugar in Peterborough where one wall had a shallow curvature with a radius of several hundred metres. Setting this out on a drawing board with compasses would have been impossible but their GDS software had no difficulty in constructing such a large radius curve. There are many other examples, particularly of suspended and cable stayed structures, which could not have been drawn by traditional means. The technology to construct them has developed over the same period but, without sophisticated CAD systems, such forms of construction might not have been possible.

4.4 Effects of IT

The effects of IT on building design are only just starting to be felt as they provide opportunities for new forms and can simulate some of their characteristics. In future, as is shown in subsequent chapters, there will be more radical effects on the whole process of commissioning buildings, pre-assembling them as models in computers, and handing them over to their users as electronic systems to be managed like a factory.

Effects of communications that are starting to change all businesses, include dispersal of companies and more international tendering of contracts. The practical obstacles to European tendering were too great before the arrival of fax and electronic mail. Companies can now collaborate from a distance, in fact the different time zones allow work to be continued around the clock in the Far East, Europe and America. Delivery and fixing of heavy building materials is still largely local but, with more valuable prefabricated components, transport costs across national boundaries are less of a limitation. The toilet pods for the Hong Kong and Shanghai bank in Hong Kong were prefabricated and delivered from Japan. Some of the steelwork for Kansai airport in Japan came from Britain. The economics of increasing the spread of tendering can have benefits such as the one quoted by the late Sir Ted Happold who, when he was at Ove Arup & Partners, was responsible for the steelwork for the Pompidou Centre in Paris. He organized its supply from Krupp in Germany to avoid the cartel which existed in the French steelwork industry.

4.5 *Marketing and analysis*

The most common type of software in use is the database management system. It forms the basis for most computerized businesses: maintaining records, storing contact addresses and generating mass mail shots which no-one now believes are personal. An early Penguin dictionary of computing gave one definition of a database as 'a file which someone wishes to seem important'. Databases are being used more and more in construction to hold records of clients, contractors and consultants for marketing purposes. Once data is collected and stored in digital form, it can be sorted in many ways and used selectively to promote a business. Architects can keep details of their clients and their particular interests. Regular clients may record the past performance of contractors, and contractors may record that of subcontractors.

> Benchmarking is a technique for raising a group of similar businesses to a common, world-class standard. The Construct IT Centre of Excellence has produced a series of benchmark reports comparing the performance of some leading firms with a world class example. These include use of IT in: *Construction Site Processes* [35], *Briefing and Design* [36], and *Facilities Management* [37]. The report on *Briefing and Design* looked at four client-led, and eleven designer-led, organizations which nominated their best projects from an IT perspective. Comparison was made with a car manufacturer representing best practice. Visits by the project team resulted in answers to questions presented on scatter graphs. Conclusions were that vertical integration within companies was more effective than integration between companies and that, in briefing, construction was well behind best practice. The Facilities Management report used the same techniques on five clients, five service organizations and a world class multinational. There was a wide variety of performance but, benchmarking is concerned with raising the level of all firms to that of the best, and IT is seen as vital to such improvement.

Managing benchmarking data tends to require computers since it should be regularly updated. Simple applications such as spreadsheets may be quite appropriate for small groups. For large databases there is often a problem in finding the important links between data. New terms such as data mining and information warehouses, combined with intelligent objects, offer a means of extracting these vital nuggets.

4.6 *The catalytic effect of IT*

Technology push has played a significant part in some of the changes taking place in construction. Where new obligations have come in, such as the CDM regulations and Quality Assurance procedures, software tools have been developed to handle the resulting documentation. Where information needs to move faster and over greater distances, communications have expanded to enable this to happen. So far the needs of construction have been met by selecting from, and adapting, the IT tools generally available, but a more focused and fundamental approach will be necessary in future, with more adaptation of general technologies and changes in construction practice.

There is a tendency to see a few spectacular developments as solving a range of problems in construction. Virtual Reality is one of these and, while its present manifestation is just another stage in the development of visualization, its potential is for simulating buildings in many ways; not just their appearance, but their performance and the process of constructing them as well.

The following chapters look at the factors which have shaped general IT systems and then the technologies which are more specific to construction. Future developments which are likely to have a more radical effect on construction are then discussed and guidelines offered for assessing their potential value.

5

Development and take up of general technologies in construction

Information technology is a general technology – it applies to all industries although their needs may appear to be very different. Hardware is the most general element with the systems software which makes it usable, while applications software and data are what make a general technology relevant to an industry like construction.

> John Sanders, formerly of Trafalgar House (now Kvaerner), when promoting electronic data interchange (EDI), asked his audience of construction professionals why their industry was slow to take up information technology. 'Under-capitalized, fragmented, project based with tight margins' were some of the responses. 'Much the same', he said, 'as the container shipping industry which has successfully introduced EDI'. But construction is still not using EDI very effectively and it will take a more universal communications system, such as the Internet, to interest a wide range of firms in applying the EDI standard messages for business transactions.

Construction is not so very special – it designs, costs, assembles and manages like many other industries. The differences are those of scale – a project is designed and built within 2–3 years and has a life cycle of 50 years or more. The components assembled tend to be greater in number and bulkier than those in manufacturing industry. Most projects are unique although prefabrication and system building could be due for a revival, if only to meet the need for public buildings for housing, education and health, which were much neglected in the period 1980 to 1997.

This chapter looks at technologies which are shared by construction, where the differences are of scale or application, but construction depends upon products which also serve a more general need.

Over the short lifetime of IT, the predominant factor affecting change has already progressed from the hardware, through the software and now lies in communications and data. These last two are the fundamentals of the information society and the first two merely the means of handling them. Computer hardware now provides small, cheap, reliable and fairly friendly devices through which we can communicate. These characteristics will continue to improve and we can even envisage input direct from the brain, certainly by voice, and display direct to the eye or, at the very least, to the end of the bulky cathode ray tube display.

> Flat screens announced at the Comdex exhibition at the end of 1996 included a 20 in (50 cm diagonal) TFT screen with 1280 × 1024 pixel resolution from Samsung and a 42 in (105 cm diagonal) Panasonic plasma screen with 640 × 480 resolution and 160 degree viewing angle. This will act as a wall-hung television and should be available by the end of 1998. The arrival of such screens was heralded in an EC Workstation study in the early 1980s and is one of the developments which have taken longer than expected. Designers of financial dealing rooms have also realized how much expensive space they could save by substituting flat screens for their many monitors. Once these screens come down to an acceptable price the shape of the PC will change, building on the growth of laptop and palmtop computers.

Hardware is now quite adequate to cope with most business needs and the software, with which this book is mainly concerned, is fairly reliable, interchangeable between most computers and, if commercially successful, well supported and continuously developed. It suffers from obsolescence and lack of integration with related applications software. **Indeed there seems to be a conspiracy of hardware and software suppliers to increase the capabilities and requirements of their systems to stimulate demand for each other's products.**

Where many future developments lie is in the organization and transfer of data; data which encapsulates all we need to know about an entity rather than just its shape, cost or performance. Object modelling techniques are likely to provide the means of doing this with their ability to encapsulate all related data and inherit attributes from other objects. Combine this with neural networks, and their techniques for learning from experience, and we are a little closer to autonomous systems.

Communications may be successful generally but, in construction, we are still near the bottom of the growth curve (see *Figure 5.1*). We can reduce most data to a digital form, combine it and send it across the world at faster and faster speeds. Where further development is needed is in conventions for interpreting and using this data, commercial guidelines, security and, as it proliferates, selection mechanisms such as knowledge agents. These are software tools for finding what we want rather than what someone else wants us to see.

5.1 Coping with change

> Alvin Toffler in *Future Shock* [38] describes the accelerating rate of change in technology giving two examples: speed of travel where, until the chariot, the limit was probably the camel at 12 kph, then, right up to the time of the stage coach, it was 30 kph. It was not until 1880 that advanced steam engines reached 150 kph, and 1938 before aeroplanes achieved

600 kph. We are now travelling in space at 25 000 kph. His second example shows the rate of change in book production, estimated at 1000 per year in Europe before the introduction of printing. By the 1950s it had reached 120 000 per year and, by the mid 1960s, 350 000 titles per year. Electronic publishing now allows even greater proliferation without ever knowing the number of readers or having to put the material onto paper.

The information society will accelerate this trend towards Vannevar Bush's boundless vision of having all human knowledge accessible through a single device (see *Figure 2.3*). Fortunately all this information can be stored at its point of origin and transmitted to where it is needed, eventually supplanting the cumbersome process of writing, printing and distribution. For the time being the construction industry's contribution to storing this knowledge, the building of libraries, is secure. In a presentation on the new British Library building, its enormous and escalating cost was estimated to be less than that of scanning and electronically storing all its contents.

Coping with change is the subject of *Future Shock*. Its author suggests 'there must be a balance, not merely between rates of change in different sectors (construction and IT for example) but between the pace of environmental change and the limited pace of human response'. Technologists see the value of their novel products as self-evident and wonder why people are slow to adopt them. For the users these products must fit into a broader business base and they need reassurance, from other users, for example. Hence there are the secure monopolies whose products are bought in large numbers: IBM in the 1980s, Microsoft in the 1990s. Who will dominate information delivery in the twenty-first century? The wealth of the media barons and the amounts that some people are prepared to invest in the media, point to where the ultimate power in the information society will lie. But power also lies with the people in that it is now possible for you or I to publish a World Wide Web page to the world at a tiny cost, if only we could then persuade others to look at it (see *Figure 2.4*). This is where different media work together and it is common to see Web addresses on advertisements or even to see advertisements aimed at promoting a Web page. Soon on-line sources will become the first place to seek information.

Future developments are not just influenced by technology but also by social and economic conditions. Francis Kinsman, in his book *Millennium* [39], simplifies these into three scenarios: **retrenchment, assertive materialism and caring autonomy.**

Since the publication of his book, each of these has had its turn. The construction industry experienced a sudden **retrenchment** in the early 1990s after an excess of **assertive materialism** in the late 1980s. The latter is returning in the mid 1990s through transfer of public services to the private sector and fewer publicly funded developments. **Caring autonomy** is a popular movement, seriously constrained by the other two, but compensating for the reduction in public support. The concept of providing a service has suffered during this period with the threat to professionalism and the rise of the product. At the beginning of the information age, professional architects, engineers and surveyors were treated as equally competent and worthy of their minimum fee scales. Since then their competence has been called into question and their services packaged to be traded in the market place. The Construction Industry Computing Association was one body set up, initially with public funds, to advise the UK construction industry and carry out various altruistic roles paid for by a supportive membership. Now its services, and those of other research and advisory bodies, must be packaged and sold like any other product.

5.2 *The dangers of prophecy*

This book is about the future and any guidelines we can derive from observation of the past. Prophecy is a dangerous business – as the old Chinese proverb is alleged to say: 'To prophesy is extremely difficult – especially with respect to the future.' The Koran is more forthright: 'He who predicts the future is a liar even if he proves to be right in the end.' More encouraging to those who attempt this necessary task are the words of Ted Kennedy: 'those who forget the past are condemned to repeat it, but those who anticipate the future are free to shape it'.

The problem in predicting the future does not lie in identifying the technologies themselves, but in estimating their rate of adoption, in this case by a complex industry subject to cycles of growth, retrenchment and occasional leanings towards caring autonomy. The Technology Foresight studies [4], by the short-lived UK Government Office of Science and Technology, used Delphi techniques to put likely timescales to a range of innovative ideas. The rounds of voting on these, of which I contributed to the studies on construction, computers and communications, concentrated the dates by which commercial success for their many ideas would arrive into the period between 1999 and 2009, although the options ranged from 1995 to 2015, or never. The implication was that, although this wide range of imaginative ideas was not quite here yet, if they did not happen within 15 years, they never would. Those of us responding to the survey by suggesting a longer time scale in the first round found ourselves in a minority and, like sheep, tended to conform to the majority view in the second round.

Methods for presenting predictions vary from the ancient oracles, which shrouded their visions in stories open to various interpretations, to more rigorous methods today. Francis Kinsman [39] suggests three approaches:

- modelling the past and using statistics to project the future;
- using imaginary people and projecting their lives into a new environment;
- using imaginary future headlines.

He chooses the last of these in *Millennium* as did the Centre for Strategic Studies in Construction at the University of Reading in *Building Britain 2001* [25]. Some of their headlines are remarkably convincing several years on, but use the effect of the horoscope writer that some are bound to be right and these we notice. Others are still desirable objectives. Examples of these headlines, written in 1988 are:

- 20% VAT on new buildings
- Ministry of housing in Brussels announces ...
- PSA privatization: five years on
- Employers win battle for common first degree
- Industry funds BRE buyout
- Crash apprentice scheme announced
- Retirement age 55 now common
- Atkins take over Wimpey-Barclays
- TGV (French high speed train) arrives in Glasgow.

In spite of their foresight no one should base important business decisions on such headlines. The second technique is that of the science fiction writer who produces a good story from a simple hypothesis, for example, the combination of a blinding astronomical phenomenon and deadly stinging plants, which nearly results in the collapse of human supremacy in John Wyndham's *The Day of the Triffids*.

This book is rash enough to follow the first path, of analysing the past to project the future, and to derive a set of guidelines for evaluation of future developments. It leaves the application of these guidelines to technologies and products of the future largely to the reader. I would be pleased to accept credit for any successes the reader may have, but will bear no responsibility for wrongful application.

5.3 *The methodology of prediction*

The approach adopted for prediction, or rather the means of providing readers with a method to assess future technologies, is based upon experience and any statistics available. It involves looking at six factors in recent technologies and relating these to data on market penetration over various periods between 1970 and 2000.

In the case of general information technologies in this chapter, the data sources are incomplete and variable and, in some cases, it is only possible to show general patterns based on my own observations. In the next chapter there is good data on computer applications in construction, largely from the 'Building on IT' series of surveys [33] of the largest 800 consultants and contractors and, for CAD, the source is the 'CICA CAD Sales surveys' [40] collecting data from CAD vendors annually from 1981 to 1995.

> These surveys were carried out by myself while at the Construction Industry Computing Association, the former with Ken France, then at KPMG, and Denis Wager, and the CAD surveys were started by David Scoins and continued by Erik Winterkorn. They provide data for the take up of applications software in their particular markets and are given as a proportion of the top 100 firms in the relevant groups in the construction industry.

Many of these technologies, particularly those involving communication, exhibit progress along an S-curve, and it is by extending this curve and its variations that we can point towards future patterns of growth or decline.

The application of this technique to more recent technologies is discussed in later chapters. One of Francis Kinsman's other approaches, the use of case studies, is also used. These are really the domain of journals, which can respond weekly or monthly to new developments, or the Internet, which can respond even more frequently. This book aims to record historical factors and help the reader project these into the future.

Technological development is made up of a series of S-curves representing succeeding technologies. In 'An IT forecast for the architectural profession' [41] Martin Betts and Stephen Oliver present the relationship, at a particular time, of existing, key and emerging technologies, each on different S-curves (*Figure 5.1*).

They also include decision support and executive information systems, two aspects of a phenomenon which has existed since computers were first considered for automated design. Can computers make our decisions for us? We humans, especially in large committees, have problems in resolving different views. If we build the evidence into a computer model, will we then believe what it indicates we should do? Prediction would be easy if we could model all future eventualities and if we had all the data on which to base accurate decisions. Unfortunately we must usually balance the predictable elements with a range of views on the possibilities. These are the scenarios used by most studies of the future.

Another technique for presentation is the four-axis diagram with the range of probabilities crossing the degree of importance axis. This allows the relationships between some aspects of different technologies to be expressed visually. The results of the survey of applications for architects by Betts and Oliver [41] are presented in this way. The most probable/important technology is shown as design visualization while the least is robots on site. The other technologies cluster around the neutral centre but on the high/high:low/low axis. (*Figure 5.2*). This does not relate to their stage of

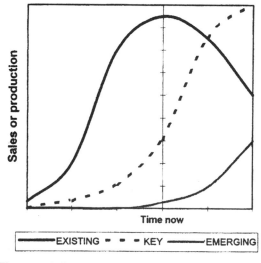

Figure 5.1 S-curve showing characteristics of technology development from a paper in *Automation in Construction* by Martin Betts and Stephen Oliver.

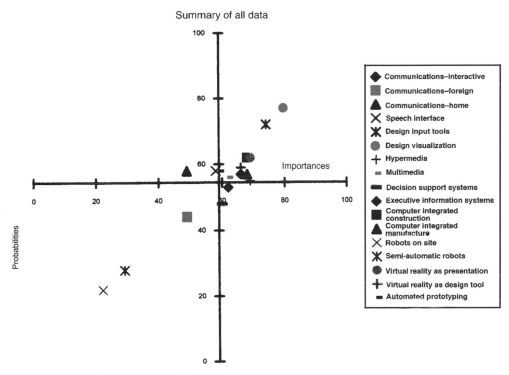

Summary of all data

Axes as means of all probabilities and importances

Figure 5.2 Probability and importance of various technologies to architects from a paper in *Automation in Construction* by Martin Betts and Stephen Oliver.

development on the S-curve but it is such indications of likely future success which those making decisions on acquiring technology need.

5.4 *Patterns of growth in general technologies*

Hardware systems – mainframe, minicomputers, personal computers and networks of technical workstations represent the range of physical sizes of system, the definitions of which have changed over time (*Figure 5.3*). The pattern of change is based on surveys of large construction firms but is typical for most technical businesses. The exception is in those maintaining very large databases, such as banks and insurance companies, which still keep large data centres with mainframe computers even if these now consist of networks of smaller machines.

Processing has become distributed to the user's PC recently but there is a move back towards centralization through network computers, completely dependent upon programs and data downloaded from networks. An assessment of the future growth of this rediscovered approach is given in Chapter 8. The main factors in the migration to smaller systems have been the reduced costs and the *de facto* standards which now exist

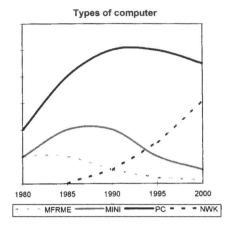

Figure 5.3 Pattern of growth of hardware technologies 1980–2000.

for general office systems.

Operating systems – CP/M was the 8-bit standard of the late 1970s/early 1980s (*Figure 5.4*). UNIX was the scientific multi-user, multi-tasking system adopted by the academic community in the early 1980s but it failed to converge on a single standard version and, although it provides the basis for the Internet, it has never conquered the business market. The proprietary Macintosh graphical user interface was introduced in the mid-1980s and set the standard for ease of use and compatibility of applications. MSDOS is the Microsoft operating system adopted by the IBM PC for 16-bit computers. This was only recently superseded by Windows95 which, with Windows NT, has become the dominant system on 32-bit computers.

The main factors affecting the changes shown in the graph have been the increasing precision of personal computers from 16- to 32-bit processing, with 64-bit systems on the way, and sales success leading to the very influential market share. Operating

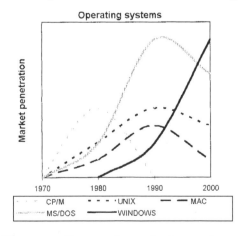

Figure 5.4 Pattern of growth of operating systems 1970–2000.

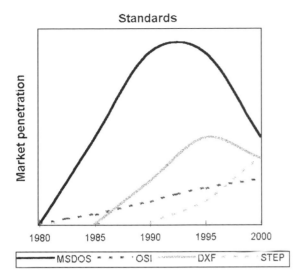

Figure 5.5 Pattern of growth of data exchange standards 1980–2000.

systems have rarely been successful as a result of their inherent 0qualities, but marketing policies and strategic partnerships have led to dominance by mediocre software
such as MSDOS.

Open systems and data exchange – MSDOS has also become a *de facto* open system
through its dominance of the PC market allowing exchange of data between Microsoft
Office products and the office automation systems of others (*Figure 5.5*). OLE, Object
Linking and Embedding, is another technique introduced by Microsoft for interlinking
data generated by different office systems. Open Systems Interconnection (OSI) is an
ambitious standards programme to formalize the interconnection of complete
computer systems using a 7-layer model to ensure compatibility from physical
connections to data structures. A large number of International Standards have been
published but the process has taken too long and has been overtaken by proprietary
products.

In the field of CAD data, DXF is the established *de facto* standard for exchange of 2D
drawing data, but it is the proprietary external data exchange format for AutoCAD
which has been implemented by most other CAD vendors as the low level solution to
data exchange. STEP, the ISO 10303 formal series of standards for exchange of product
model data for all industries, is a more comprehensive modelling standard but is again
taking much time to complete and is likely to materialize for construction through
more focused initiatives such as the International Alliance for Interoperability.

**Information technology standards consist of long-term formal standards which
are often superseded before they are widely adopted. These may influence more
pragmatic *de facto* standards which may succeed more quickly if associated with
products dominating a particular market.**

5.5 *The factors which affect success*

In Figures 5.6–5.10, a number of factors have been selected as influencing success, but these cannot be defined precisely for the whole range of technologies to which they are applied. The following are general descriptions:

Application identified – this factor indicates when the use of information technology, for this purpose, was established. For example there has always been a need to edit and print text so that the demand for word processing was well established before computers could do it efficiently. Spreadsheet programs are different in that the idea of a table which recalculates when any number is changed, was relatively new when the first such software appeared in about 1980. Such a facility met an unspecified need and quickly grew to saturation level among business users.

Hardware adequacy – this was the aspect of information technology which advanced most rapidly over the period 1970–1990 and there are several elements of performance which influenced the success of its use for particular applications: speed, memory, storage capacity, display technology, reliability and ease of use. All of these elements had to be adequate before the application could achieve success.

Software packaging – in a market looking for reliable systems, it is important to distinguish between packaged software which is supported and maintained, and the situation in the 1970s when most computer users were programmers and would produce routines, when they were needed, for particular tasks. As time progresses it becomes more dangerous to change software on which a business relies unless it is to use the built-in flexibility offered by databases and Fourth Generation Languages. Most software began as a specific solution for a single company or research project. Software is now big business and success is dependent upon conforming to standards, ease of use, support, regular updates and market share itself.

Market established – general purpose software has an enormous market across all business and cultural activities. The need for IT systems in all these areas is established through education, publicity and the efforts of companies promoting products. Success is achieved with the bundling of software with hardware or its availability as shareware or freeware. Other applications, such as those specifically for construction, address particular niche markets and may quickly become established in these. This factor measures each application against its own market, although even industry specific software does migrate from one market to another.

Benefit exceeds cost – this is difficult to generalize for all users since what is measured as a benefit to each can vary. Many applications are highly priced initially to recover development costs and even to confer status on the product. If successful and, if other factors are right, costs may reduce as the market grows. If critical mass is reached, the S-curve is typically followed until superseded by another product. Very rarely, in the

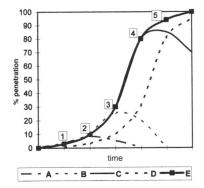

Figure 5.6 Typical family of S-curves showing the growth of technology.

short history of software, has the basic price increased except when there are substantial enhancements. This might change as particular products start to dominate a market and software piracy eats into profits.

Standards in place – this last factor usually follows the establishment of a new technology. Standards, according to the official standards bodies, are meant to reflect common practice. In IT this would be too late and future developments must be anticipated. Confidence is given if applications software uses the most popular hardware or operating system – these become *de facto* standards and are often more influential than the formal ones. It is usually the relationship between linked software packages and the data they may need to exchange which requires standards, preferably formal ones, if they can be produced quickly enough. A new product which is very successful can become a *de facto* standard in its own right.

5.6 Stages of growth

On the typical S-curve graph of applications developing over time (*Figure 5.6*) the following points and alternative paths to the successful system are significant:

1. First trials   A. Ideas not reaching the market or failing quickly
2. Market opens   B. Decline after some success and being superseded
3. Critical mass   C. Decline after wide usage – inevitable at some time
4. Well established   D. The succeeding technology
5. Saturation   E. The typical S-curve of the successful technology

5.7 General applications used in construction

The data used in the line graphs is based upon surveys of large construction users [33] and presents trends based on their expectations for the near future. Significant factors are presented in the histograms on a scale of 1–5. Individual technologies of

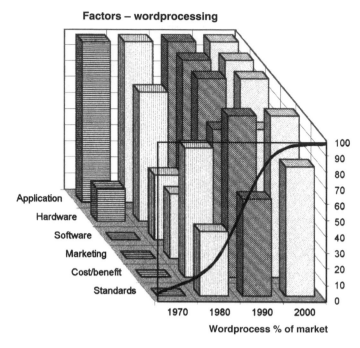

Figure 5.7 Word processing. Pattern of growth and influential factors. 1970–2000.

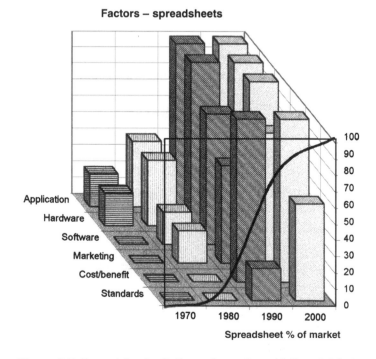

Figure 5.8 Spreadsheets. Pattern of growth and influential factors. 1970–2000.

particular significance for construction are projected into the future in Chapter 8 by applying the factors which appear to be particularly influential in their success.

Word processing – this exhibits a typical S-curve to meet a very widespread need which has always existed (*Figure 5.7*). It replaces a typing function. The hardware has been suitable since dedicated word processors, Apple Macintosh and PCs have been available. The number of software packages is now converging on a smaller number of comprehensive systems. Marketing has been strong and cost/benefits have been self-evident for repeatedly edited, or widely issued, documents. Standards are very basic, ASCII character sets and Postscript for printers, but systems have generally allowed data conversion for other popular word processors.

Spreadsheets – again very widely applicable but they introduced a new application when they appeared on PC (*Figure 5.8*). Very little financial modelling was carried out until then and this is one application which has transformed business planning since Visicalc was launched around 1980. There have since been a range of excellent packages. Marketing was strong in the 1980s, largely to sell PCs, and Lotus 123 and Microsoft Excel, available for both Macintosh and PC, became market leaders. Benefits are not hard to prove but standards are limited to conversion of data to the format of other common systems.

The contrast with word processing is that a new application was effectively created and had to generate a new market, which it has done with great success, and followed an even steeper S-curve than word processing.

The Internet – is a very recent phenomenon where the only precedents were proprietary networks most of which now link to the Internet while trying to offer added value (*Figure 5.9*). This phenomenal growth is charted on a logarithmic scale derived from data published by the Internet Society [42] and is based upon the number of hosts or individual information providers. Their prediction is for 187 million hosts by the year 2000. This is one for every 32 people in a projected world population of 6100 million. The OECD prediction is slightly lower at 142 million users by 2000. See their Web site at *www.oecd.org/sge/au/highligh.htm*

The factors affecting this amazing growth have been the latent application for global communications, the realization of which at low cost, has only recently become possible. The technology of communications networks has become more important than software or hardware, although fast modems, to connect computers to telecom lines, are necessary for World Wide Web data. Standards were the major factor and here the work of the CCITT, the standards body for telecoms, has been much swifter and more effective than ISO working in other areas. The TCP/IP protocol is a *de facto* standard on which the Internet is based, while Microsoft's ActiveX and Sun's Java are languages competing to dominate Web authoring.

Computer-aided design – is a general technology although the graph (*Figure 5.10*) is based upon the systems most widely used in construction from the annual CICA CAD

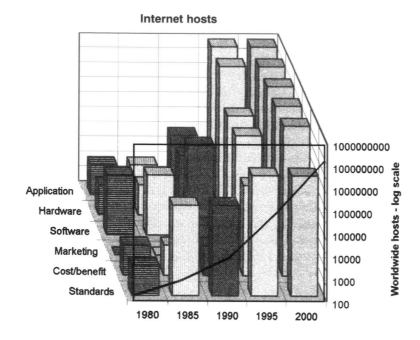

Figure 5.9 The Internet. Pattern of growth (log scale) and influential factors 1980–2000. The Internet Society.

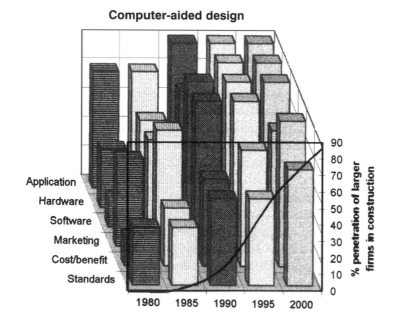

Figure 5.10 Computer-aided design. Pattern of growth and influential factors 1980–2000. CICA annual CAD sales surveys and projections.

systems sales survey [40]. A more detailed analysis of use by the different groups in construction is given in the next chapter. Market share for all industries is more usually quoted by value of turnover in surveys by Daratech or Frost & Sullivan. More relevant to such technology changing the way an industry works, is the number of firms which have adopted it, and the graph is based upon the proportion of potential user firms. Continuing growth can be expected in the number of workstations in a design office and from the tendency towards having multiple CAD systems to suit different projects.

The dominant factors are not the application which, for draughting, has always been a major element in building design, but the capacity and speed of the hardware for handling the large amounts of data, and the ease of use and efficiency of the software. Marketing has been strong and costs and benefits more demonstrable with the rise of CAD systems on personal computers.

5.8 Summary

The most widespread uses of computers in construction are for quite general technologies but the most strategic ones are specific to the industry. Evolution is from the general to the particular and, as those in construction have become more familiar with software, it has been tailored to meet their specific needs. However, people are still trying to catch up with the possibilities of technology and the IT industry appears to make this difficult for them. The conspiracy between hardware and software suppliers to create demand for the latest versions of each others' products forces users to invest frequently. More change results from the opportunities offered by new technologies than from feedback on users' needs.

A variety of methods of prediction are used to try to anticipate the future but new technologies often arrive unexpectedly. The Internet existed for years as an academic service unnoticed by the business community and was suddenly discovered. Few pundits predicted its phenomenal rate of growth.

There are dangers in prediction, but there are also signs which can be used to assess new technologies. The factors which have already influenced the success of some general systems and applications offer some guidance. Those in construction can learn from these factors but will be most influenced by the factors affecting take up of construction specific applications, the subject of the next chapter.

Major contributions to successful application in building

6.1 The evolution of computer applications in construction

Before analysing the characteristics of the applications software which has, and has not been, successful in construction, it is important to consider the environment in which it emerged, the organizations which helped to encourage usage and the inertia of a large and complex industry with many parts.

> The growth of applications resulted from the arrival of programmable calculators and time-sharing computer bureaux in the early 1970s. I was fortunate to be working with Applied Research of Cambridge (ARC) in this formative period. The big system building programmes described in Chapter 3 were spinning off general purpose applications for computer-aided design and analysis, and the fruits of university research were starting to become available as clever, but unreliable, software. The Design Office Consortium was formed as an alliance of ARC and the Department of Trade and Industry CAD Centre, operator of an Atlas computer offering one of the first graphics time sharing services in the UK. DOC, which became the Construction Industry Computing Association in 1980, was established to promote and support the first applications packages.

A bureau service was offered on Atlas for a range of types of software: the Cambridge Environmental Package, design generation and appraisal software from the ABACUS unit at Strathclyde University, visualization software from the CAD Centre and earthwork excavation analysis from ARC. Use of these systems was not easy with little data verification and slow, unreliable remote access using acoustic couplers, the forerunner of the modem. There were also limited computer resources; 128 Kbytes of memory and 24 Mbytes of storage were shared between 20–30 users on Atlas (*Figure 6.1*). (Compare this with the size of the hand-held Apple Newton with 1 Mbyte of memory 20 years later.) DOC's role was to prepare data from drawings, run the programs and interpret the results. It carried out cut and fill analysis of earthwork

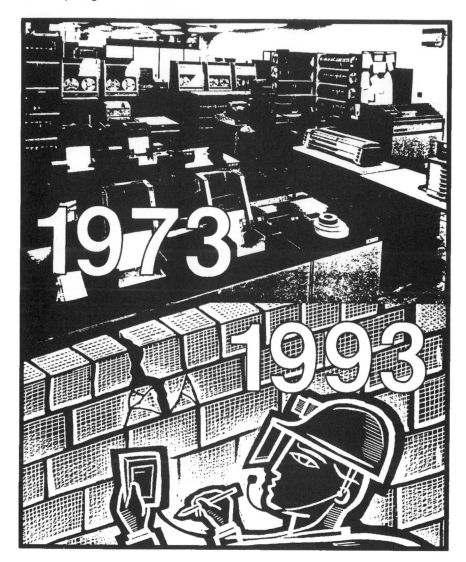

Figure 6.1 The Atlas computer at the CAD Centre contrasted with an image of a hand held data capture device from *Building* magazine to reflect 20 years of CICA.

quantities for Hampshire County Council school playing fields, visualization of Crown Court buildings for the Property Services Agency, and environmental analysis for building services consultants.

In the mid-1970s the minicomputer arrived and some larger companies could afford to buy their own. There were also desktop programmable computers on which routine structural calculations could be carried out. As the number of applications grew and more software houses specializing in construction appeared, most of them formed by people from the industry based on their research and experience, so DOC's role changed to listing available software and, since its quality was variable, to evaluating it and publishing the results for the DOE Property Services Agency. Mike Chaplin, a surveyor with PSA, became chairman and a series of respected, impartial reports was published and compared the performance of similar programs with each other or with alternative means of calculation. These reports gave users confidence to try the software and, when microcomputers appeared in the late 1970s, the stage was set for rapid growth of construction applications.

Several organizations were formed to aid the exchange of computing experience, both in the UK and overseas. Professional institutions set up computer committees but, in some cases, they were mainly to ensure that their members did not lose ground to other professions. The European Community provided research funds for two studies of computers in construction, the first led by the CIAD group in the Netherlands with W.S. Atkins as the UK partner. The second led by CICA (as DOC had now become) with partners at I3P in the Netherlands and the Technical University of Munich [66]. The subject of the second was to specify a workstation for the construction industry looking about five years ahead, and a parallel study, led by Wolfgang Haas of RIB Stuttgart, was on data management. The result of the study led by CICA was a series of three modular workstation designs based on the new UNIX operating system (*Figure 6.2*). The emergence of the PERQ and Whitechapel low-cost workstations about five years later proved that the specification was about right even if their prices turned out to be much lower than predicted.

Such collaboration led to the founding of Working Commission No 78, Integrated CAD, of the international research group, CIB. I organized a meeting at the Building

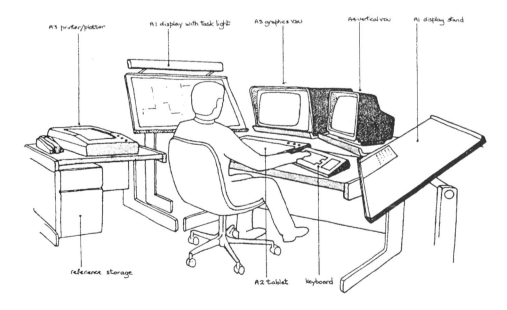

Figure 6.2 Designs for a CAD workstation from 'The specification of a building industry computer workstation' for EC DGXIII by CICA, 13P and the Technical University of Munich 1982.

Research Establishment in 1984 and it attracted a gathering of leading international researchers. Some perceptive topics were selected for future discussion:

- Sensitivity of design appraisal factors.
- Classification systems related to CAD and layering.
- Standards – particularly IGES/AEC and translators available.
- Systems for transfer of quantities from design to construct.
- Japanese fifth generation – logic programming, extensible languages.
- Relationships with other CIB working groups.
- Collecting data on systems and their usage.
- Quality assurance and source labelling of programs and data.
- Remote databases and external libraries.

The group has continued to meet regularly, often with other CIB working groups such as W74 Information Co-ordination. It also maintains a news group on the Internet.

Another international group formed at this time by user associations to share their knowledge of software and sell publications, was FACE, the Federation of Associations of Computers in Architecture, Engineering and related fields. Its main supporters were Gerard Kruisman of CIAD in the Netherlands, Martin Jones and Ian Harvey of ACADS in Australia, and Joe Rogers and Pat Johnson of CEPA in the USA, with CICA as the UK representative. Its constitution was enshrined in Dutch law and it was launched in style at the New York Hilton Hotel in 1981. However, its restrictions on members being non-profit distributing and not owning software, were such that few organizations in other countries were sufficiently impartial to be admitted although at various times it had members in Denmark, Finland, South Africa, Japan, Canada and Israel. Through these organizations, and the growing number of international conferences, experience of the application of computers in construction spread rapidly during the 1980s.

The environment of new computing technology started to increase the exchange of experience between different professions. It was recognized that the traditional boundaries had to be re-examined in the light of computers and, particularly, communications. Institutions were demonopolized in the UK at this time and it was through groups such as the CICA Advisory Committee, on which the leading institutions and research bodies sat, that there was some co-ordination. Professor Ted Happold of the University of Bath chaired this group and went on to found the Construction Industry Council, which helped to unite the professions and led to the formation of the Construction Industry Board, now a single voice for the industry.

The transfer of data throughout the project, from the computers of designers through those of surveyors and contractors, as portrayed in the 1982 film, 'Building on IT', became more feasible. This film was produced by Maiden Films for the Builder Group with support from the Department of the Environment and others. It was the construction industry's contribution to IT Year and was launched at the London Building Exhibition. The theme was exchanging data between the computers of different members of the building team. Contributors included Andrew Chadwick as the architect designing a house with CAD, Martin Hawkins, then at Cyril Sweett & Partners, as the QS taking off quantities with a digitizer, and John Hollingworth of Wimpey pricing the bills on their mainframe. Unfortunately data transfer at that time was on paper, albeit generated by computers, but the main point was that all this digital data should be transferred directly between the computers of the project team. Communications by satellite were presented by Ove Arup & Partners, who provided the location for most of the filming, with the London Telecom Tower in the background, symbolizing its use for transferring drawings to their Hong Kong office. The film ended with the message that electronic transfer of data would become the norm, and the presenter walked off carrying the 'office of the future', an Osborne 2 portable computer the size of a large suitcase with a tiny 15 cm screen. The 1982 hardware shown now looks very dated but the messages, about altering the way the industry works to enable data to be exchanged between the project team, are as fresh as ever.

6.2 Recent research reports

A number of research studies in the 1990s have shown how data transfer between different types of computer system might come about and where it is likely to lead:

Construct IT – bridging the gap [43] was sponsored by BT and published by DOE. Its aims were to promote the use of current technology and the integrated project database and to undertake detailed analysis of research and development needs. It produced an IT vision for the future, assessed current uses of IT and the barriers to achieving the vision, and offered guidelines for moving towards that vision. The vision had three elements: an integrated project communications framework, industry-wide information and specific improvements to the process. There was some criticism for repeating old themes and ignoring past experience but the projects that have followed, particularly on benchmarking [35, 36, 37], have been very successful and the DOE research strategy has been redirected towards more important themes.

Technology Foresight [4] was a cabinet office initiative for all industries and looked 25 years ahead. Its report for construction, based on the Delphi technique of using the opinions of a variety of experts, contained much IT content but the main recommendations were more general. They included setting up learning networks, better distribution of information, encouragement of long term investment and an innovative culture. The results of the survey on likely timescales for particular technologies were not published, but the opportunities to be exploited by public-led organizations included: customized solutions from standard components, construction to last for only as long as required, better assessment of environmental and social consequences, and a competitive infrastructure.

Figure 6.3 Sir George Young, Minister for Housing, welcoming the formation of the Construction IT Forum at the Interbuild exhibition in 1995 with, on left, Phillip Ward, Director DOE Construction Sponsorship Directorate, and the author and, on right, John Trussler, Chairman of the Forum and Graham Watts, Chief Executive of the Construction Industry Council.

Building IT 2000 and Building IT 2005 [5] by the Construction IT Forum looked ten years ahead to those dates respectively. They were collections of views by experts in a range of construction and IT fields and assessed the current state of technology, the possibilities ten years on, the inhibiting factors and the actions needed to minimize these. The second of these was distributed as a multimedia CD-ROM and conclusions were aimed at three groups: clients, the UK government and other countries; the building and construction industries; and the IT industry. They were largely intended to promote industry take up of existing technology, further development of this, and specified some short- and long-term research needs. Hopes were expressed for the implementation of these by industry encouraged by the Construction Industry Board and its CRISP research panel.

With such a variety of reports appearing in 1995 there should be little need for further exploration of the future. However, as far back as 1967, the Minister of Public Buildings and Works, Reg Prentice, said that 'the time would come when design and construction information would be passed as readily between one group and another as a telephone conversation took place today' [67]. Each of these recent reports tends to reinforce the messages in the others even if the Construct IT report tends towards the idealistic, the Technology Foresight studies are more socially based with little on specific technologies, and the Building IT reports have suffered from the failure to continue the work of the Construction IT Forum (*Figure 6.3*). It seemed possible at one time that the expertise behind each study would be incorporated in the new CRISP research initiative. There was a great desire to unify the voice of the construction industry and this is now being done by the Construction Industry Board.

It is against the background of the expert and widely consulted views in these reports, and the changing environment summarized at the beginning of this chapter, that the success of individual applications can be assessed.

6.3 Draughting and modelling

Most building information is in the form of drawings and a graphical capability was essential for applying computers. This arrived with the storage tube, for displaying graphics, in the mid 1970s. Pen plotters had been available for some time but had generally been used for graph plotting only.

Three applications were identified within what is now called Computer Aided Design: *drawing production* based on automating 2D manual drawing techniques, *3D modelling* from which 2D drawings and perspectives were forms of output, and *automated design*.

Researchers are quick to spot the potential of new technology and the last of these applications was an early topic for investigation. Computers could only address a few of the many factors which need to be taken into account in design, and designers were generally satisfied with the methods they were already using.

Draughting was likely to be the most productive since 30% of an architect's time was estimated as being spent at the drawing board, and there is much repetition on larger projects and frequent design changes.

Modelling involves describing a building in 3D and can include attribute and performance, but little of the data to describe components fully is available at the early design stage. The external form could be modelled relatively easily and perspective drawing programs allowed multiple views to be generated. These were used initially as a basis for accurate, but manually rendered, presentation drawings.

As the number of available CAD packages grew in the late 1970s and 1980s, the early ambitions for complete 3D modelling and automated design were put aside and drawing production became the most realistic goal of architectural, and later, engineering, consultants. The graph (*Figure 6.4*) shows the number of packages available, cumulatively, so there were very few draughting packages in 1975 but, as the number of these grew, the total number of firms using them also grew following a typical S-curve. By 1996 this was approaching saturation in architectural and

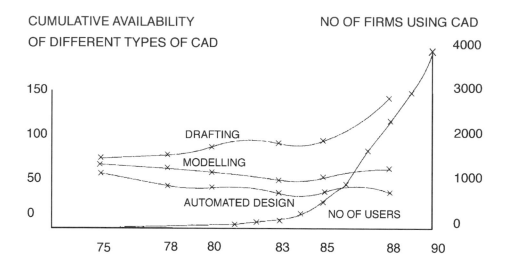

Figure 6.4 Cumulative availability of types of CAD system and growth of user firms in construction. 1975–1990. CICA software directory and sales surveys.

engineering firms employing more than 10 people but the number of workstations deployed is still growing and there is much expansion yet to come from smaller design offices and from design and build contractors.

6.4 Take up of CAD by different groups

Data on the growth in the use of CAD by different groups in construction is again shown as a cumulative graph (*Figure 6.5*) and is based on the annual CICA CAD Sales Surveys [40]. A projection of the number of workstations in different types of firm is given up to the year 2000. This is based upon data from a selection of 10 major CAD system suppliers. Factors in this growth are, of course, the reducing cost of CAD hardware and the increasing reliability and ease of use of systems. In 1980 a single seat CAD system on a minicomputer cost about £100 000. By 1985 the IBM PC had arrived and relatively simple CAD systems were available for £20 000. By 1995 the capabilities of the most powerful systems had been provided on fast PCs at a price of £3000–£5000.

Research and Teaching establishments are relatively small in number but now have substantial numbers of CAD workstations and also provide training for businesses.
Multiprofessional firms have not grown in number with the reduction of public design offices, but mergers of architects and engineers have created new firms which are often large and committed users of CAD and have the most opportunity for benefit.
Consulting Engineers took up CAD rather later than architects but their other engineering analyses have naturally extended to the addition of graphics and CAD.

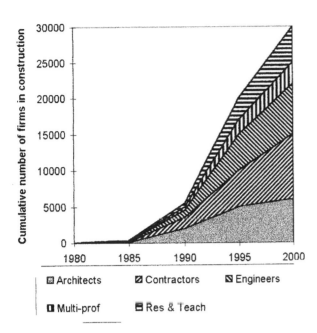

Figure 6.5 Number of CAD user firms by type 1980–2000. CICA CAD sales surveys.

Contractors show continuing growth past the end of the century since firms other than the largest started to use CAD very late, and drawing has only recently become an essential application in housebuilders and those offering design and build.
Architects, most of whose work is graphical, were the first to explore CAD and to use it for visualization. By 1994, 60% of firms already had at least one CAD system [64].

Growth in the number of workstations is likely to continue upwards until there is one for each designer, and firms may have several different software packages as is already common with Macintosh users. Surveyors, particularly Land Surveyors, are also CAD users with about 20% of larger firms reporting current use in 1993. Housebuilders indicated about 80% of the larger ones had some CAD capability in the same survey, *Building on IT for Quality* [33].

6.5 *Development of communications in construction*

During the 1990s the most significant change in information technology was in communications. In the 1980s a few leading firms experimented with exchange of project data on fixed media such as disc or tape, but this usually meant transfer by courier. By 1995 direct electronic transfer over telecom lines had became reliable.

The last *Building on IT for Quality* [33] survey of large construction firms in 1993 did not even include the Internet and yet, by 1996, a significant number of firms in construction were connected; 25% of UK architects were connected in 1997 with a further 15% intending to open an account [62]. Many contractors are wary of linking to their internal networks for reasons of security, although they use the same technology in the form of Intranets for their internal communications. The future use of the Internet is likely to be huge as indicated in the graph (see *Figure 5.9*).

The types of communications surveyed in 1993, with future expectations from all sectors of construction (*Figure 6.6*), were:

Electronic Mail then largely confined to internal networks, now globally by Internet.
Electronic Data Interchange (EDI) for commercial messages which, for construction, have been standardized by EDICON in the UK and EDIFACT internationally.
On-line databases which were largely confined to research and bibliography before the arrival of the World Wide Web, with some trials of building product data.
Compact Disk – Read Only Memory (CD-ROM) disk drives used for loading complex software and accessing the growing amount of data presented in multimedia form.
Integrated Services Digital Network (ISDN) providing high bandwidth telecom lines suitable for transferring large files, such as drawings, rapidly.

Expectations were high for future growth of these services, and indications given in Chapter 8 are based on these and the developments in communications technology. Email has increased greatly due to the rise of the Internet as has the use of on-line databases. EDI will be stimulated by this technology and CD-ROM is enjoying a period of stability before the new Digital Video Disc (DVD) standard arrives with a

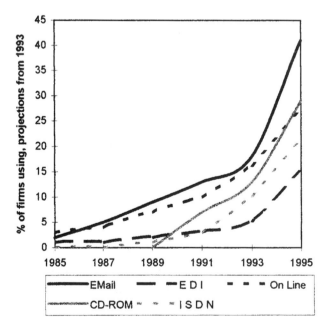

Figure 6.6 Growth of communications technology in large firms in construction 1985–1995. *Building on IT for Quality*. CICA, KPMG. 1993.

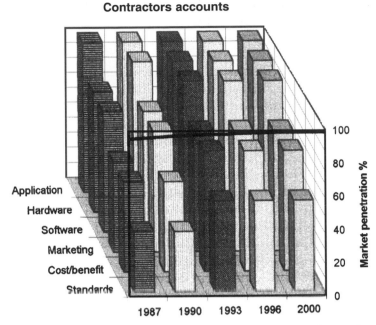

Contractors accounts

Figure 6.7 Accounting systems. Growth of firms using software, future projections and influential factors 1987–2000. *Building on IT for Quality.* CICA, KPMG. 1993.

capacity of at least 6 times that of CD-ROM. ISDN lines are still expensive to install in the UK but their ability to handle the growing amount of multimedia data will lead to their becoming the standard means of business communication with cable systems serving a growing number of homes.

6.6 *Contractors' accounts and estimating*

Just as CAD is the basic application of the designer, integrated accounting systems fulfil this role for the contractor. Major contractors were already running their ledgers on service bureaux or mainframes in the 1970s and have moved this strategic application onto minicomputers, and now, networks of microcomputers, adding more modules for payroll, subcontractors, contract costing, plant hire and cash flow.

Accounting, as the data from a sample of the top 400 contractors shows (*Figure 6.7*), has been in continuous use from the 1980s with saturation among the larger firms. Surveys by the FMB and CIOB indicate growing levels of use by smaller firms. The need has always been there and hardware and software have become more capable, providing increased cost/benefits. Standards depend on accounting standards set by auditors. Modular systems tend to work only with those from the same source.

The major factor with accounting systems is the need for fast, consolidated financial information from all projects. Management without such systems is now very difficult.

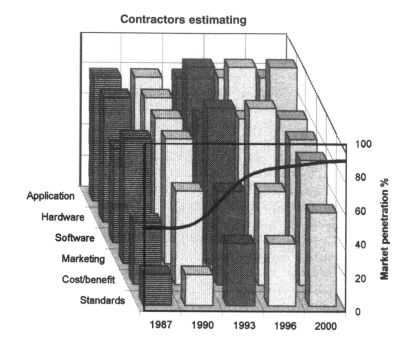

Figure 6.8 Estimating systems. Growth of firms using software, future projections and influential factors. 1987–2000. *Building on IT for Quality*. CICA, KPMG. 1993.

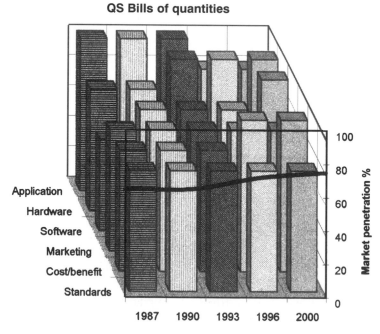

Figure 6.9 Bills of quantities. Growth of time using software, future projections and influential factors 1987–2000. *Building on IT for Quality.* CICA, KPMG. 1993.

Estimating by computer has long been desired but it has taken some time for the many available software packages to be used to their full capabilities. Estimating is a very sensitive aspect of contracting and is regarded by many as an art. However, the pricing of quantities can be carried out as a data processing exercise and the weightings added by experienced estimators. It has taken many years for some estimators to see the advantage of being able to look at options, but the application is now used in most large contractors (*Figure 6.8*).

Marketing of estimating systems has been strong but the benefits have only recently been sufficient to convince sceptics. With a larger share of work going to subcontractors, electronic communication of their prices will help main contractors to consolidate their tenders, and sales of estimating systems should continue to grow.

6.7 Bills of quantities and project management

For Quantity Surveyors the bulk of their work has traditionally been in measuring, working up and printing, complex bills of quantities according to Standard Methods of Measurement. This was an obvious application for computers and usage started in the 1960s. Fewer bills are produced now and the QS provides early cost advice using cost planning packages and spreadsheets and often manages projects with the many networking tools available. Contractors and Civil Engineers are probably the biggest users of the many bar charting and critical path programs on the market, but the surveys show that there is continuing potential for growth in this area.

Bills of quantities is a well established application and even the early mainframes were used as printing machines. With cheaper hardware has come greater interaction and digitizers aid taking off quantities from drawings. The threat always exists of the bulk of quantities being produced from CAD models but, while the Standard Method of Measurement prevails, there will be many details which are not drawn but must be measured. The number of large surveyors using computers for bills remains below 80% but not all firms produce SMM bills of quantities (*Figure 6.9*).

Project management packages are shared with other industries, although those with linked resources are designed to handle data on labour and materials. This is another early application which the construction industry has adopted rather slowly. It is only recently, following periods of rapid inflation and the demands of clients for greater control, that bar chart or network output has been regularly updated during a project rather than being plotted once just to comply with the contract (*Figure 6.10*).

The reducing size of hardware has made it possible to take portable machines onto the site to record progress. This has stimulated use of project management software, of which there is an enormous variety. Integration of networking with complete financial control in systems, like Artemis or Primavera, has provided very powerful planning

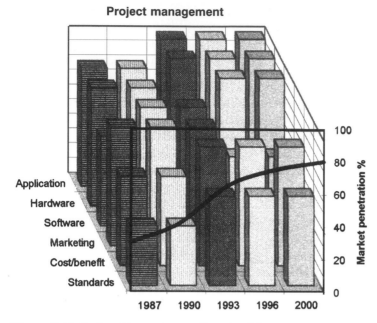

Figure 6.10 Project Management. Growth of firms using software, future projections and influential factors 1987–2000. *Building on IT for Quality.* CICA, KPMG. 1993.

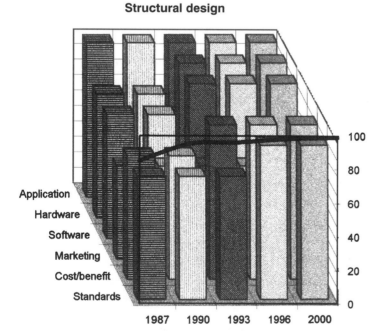

Figure 6.11 Structural design. Growth of firms using software, future projections and influential factors. 1987–2000. *Building on IT for Quality.* CICA, KPMG. 1993.

tools. The application, which started with the space race, is now providing the construction industry with a tool to control its performance.

6.8 Engineering applications

Within the two main branches of construction engineering, the specialist applications are structural analysis and design for the structural engineer, and environmental analysis and building services design for the building services consultant or contractor. These are based on established calculation methods although codes of practice are revised from time to time.

Structural design whether in concrete, steel or other materials, is invariably carried out by computer today and certain applications, such as finite element analysis, could not be done any other way. The market has been supplied by a small number of software houses, usually attached to engineering consultants, and all large firms have had computer systems for many years. They need regular updating particularly to keep up with the change from elastic to plastic codes, and now, in Europe, to Eurocodes. Analysis is more international with the same systems sold round the world *(Figure 6.11)*.

Services design is similarly supplied by a small number of software houses in the UK although there are environmental analysis models, typically from US Federal agencies, which are available worldwide. Environmental analysis and the design of energy efficient service installations has been stimulated by the various energy concerns: first the rise in the price of oil in the 1970s, and then the growing concern with global warming. Hardware developments have allowed more complex models to be built with more frequent calculation based on the dynamics of continuous climatic data. Links with CAD systems allow results to be visualized and service layouts to be added to drawings. Standards are based on published methods of calculation but progress towards European or international methods is slower than with structures *(Figure 6.12)*.

Engineering applications have not been marketed as strongly as some others probably because engineers are more familiar with calculation methods and know that they must have these programs to carry out their projects. There is likely to be more growth amongst services contractors. Control systems are becoming more complex and involve a whole range of networked building management systems.

6.9 Summary

Just a few of the applications specific to construction have been mentioned. There are 112 headings in the CICA *Software Directory* [44] and, of these, some 85 are specific to construction or contain software which has been customized to suit the needs of building design, construction or management. Most of these applications were explored

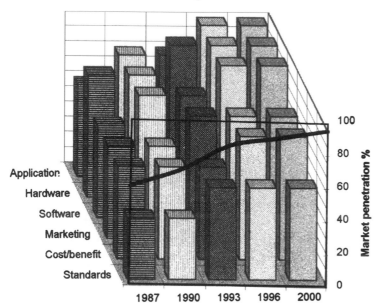

Services design

Figure 6.12 Services design. Growth of firms using software, future projections and influential factors. 1987–2000. *Building on IT for quality.* CICA, KPMG. 1993.

early in the history of computing, often as part of the integrated building systems on which the first major investments were made. Once confidence was established in the methods used and the hardware became accessible to all, the marketeers could take over. Confidence was further established by sharing experience internationally, even if much software was localized, although the dominance of the USA in computing generally has meant that most other countries tended to follow their lead.

When the integrated systems development programmes, and their public funding in the UK, declined in the late 1970s, it became necessary to promote the idea of integrating the individual applications. The idea of design automation was abandoned in favour of drawing production, and geometry was seen as the basis for integration for designers, even if money and time were still the integrating factors for contractors. As telecom systems have improved, the transfer of data has become routine and the use of electronic mail has started to follow the S-curve upwards, followed by the use of CD-ROM particularly for distributing ever larger software packages (see *Figure 6.6*).

Routine applications have become established in many areas, and are essential for engineering design and financial management. The main factors affecting their growth were: meeting an existing need, the rapid improvements in hardware and software, marketing of the most prominent packages and an improving cost/benefit ratio.

The next chapter looks at factors affecting the success of newer and more radical technologies.

7

Conditions for success in the construction market

Very few new developments in information technology prove to be dead ends. This is unlike the patents that have been registered for all sorts of ingenious devices over the centuries, very few of which, like perpetual motion devices, come to fruition (see Figure 2.1). The closest equivalent to perpetual motion in IT is the progress towards infinitely fast transfer of vast amounts of data and this is gradually being approached but through evolution rather than sudden inspiration. Most new developments of hardware or software are extensions of existing systems or combinations of IT applications to address new tasks. Few are immediately successful and this book is concerned with identifying the characteristics of those which are likely to be of significant value to the construction industry.

As well as addressing the needs of users in the construction industry, it should also be of help to those developing IT products for the industry. Collaboration between both these parties is essential for achieving the full benefit of IT. In launching new products or services, the two main factors in success, apart from applying IT to tasks which are performed regularly or with difficulty, are timeliness and having the confidence and resources to mount a marketing campaign of sufficient size while being prepared to wait for the benefits.

7.1 *The right time*

People have fertile imaginations and, almost as soon as computing became established, some saw what it should be capable of achieving. One only has to look at science fiction or comics to see images of information systems of the future, robots and display screens and an assumption of boundless communications. In construction the concept of a single, comprehensive building model was an early target even when computers were capable only of handling the crudest representation. Automated design soon became the subject of much academic research which did not meet any need expressed by designers. New techniques are starting to make this possible but it has required further progress through the IT development cycle to make these targets achievable during the 1990s.

In order to place future developments in context it is important to realize the stage we have reached in the evolution of IT systems. The development cycle has given priority, in turn, to the main elements within an IT system: *hardware, software, communications and data*. Until the first of these gained sufficient speed, capacity and robustness, the succeeding elements could not make progress. Improvements in software, using the greater speed and capacity of hardware, allowed more understandable interaction and software which non computer experts could use. Communications improved with better telecom lines and standards. All these developments require good, digital data on which to operate.

- The 1970s was the decade of hardware with development from timesharing mainframes to the first microcomputers.
- The 1980s saw emphasis transferred to software, with the emergence of standard operating systems and availability of most applications software for construction.
- The 1990s is the decade of communications with the Internet spreading out from the academic community, initially into business where most companies now have Email addresses, and into many homes via TV set top boxes by 2000.
- With all these elements of technology in place, the next century should see true value given to the data on which all systems depend. The best data will be expensive and limited to subscribers or subsidized by advertising, but the income generated will ensure its quality and regular revision.

In the early days of computing data was assumed to be the free element within a system, or that which was generated to solve the particular problem in hand. As systems have become capable of storing and processing vast quantities of data, its importance has grown and capture of data is becoming the most vital component of any system. Businesses can gain an edge over their competitors by using data mining techniques to learn more about their customers. Loyalty cards offered by chain stores may give a discount to the customer but they are primarily a means of analysing customers' preferences and responding to these.

In the UK construction industry of the 1970s, when government took a paternal view, the DOE Standing Committee on Computing and Data Co-ordination [23] identified the components of the data bases needed by the industry. Inevitably, having analysed the problem well (*Figure 7.1*), they left the solution to an industry which was just starting to be interested in acquiring hardware and trying out software on routine tasks. By the 1990s most of the pieces of the industry data base are in position. RIBA Companies have made a major contribution with the National Building Specification, Specification Manager, RIBACad and Annotation Manager, as well as being one contributor, with Barbour Index, On Demand Information, Hutton & Rostron and Technical Indexes, to electronic libraries

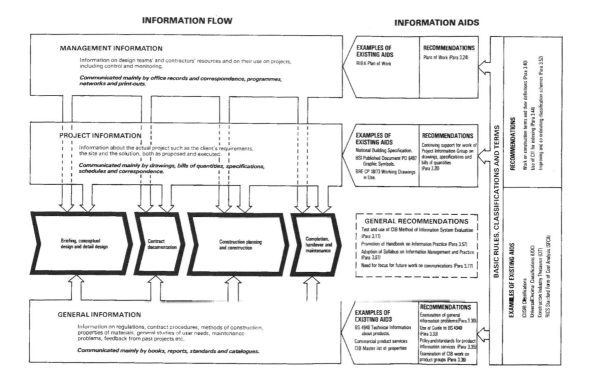

INFORMATION FLOW

INFORMATION AIDS

Figure 7.1 Communications framework proposed in the National Consultative Council report on 'Computing and communications in the construction industry'. DOE.

of building and product information. The RICS, through the Building Cost Information Service, has provided on-line cost information and the engineering institutions, the ICE and CIBSE, have played their part with the New Engineering Contract and the CIBSE Guide calculation routines. The industry generally does reach many of the goals set for it eventually, even if the solutions come from private enterprise and are sometimes diverse and incompatible. **The real need is for reliable and authoritative data in a form which can be delivered electronically and this is now becoming an economic proposition.**

These information services address the commonly used elements of building documentation. Project data is the province of individual project teams but is largely gathered from common information resources. The Project Information Group, supported by the main professional institutions and DOE, showed how project data could be organized better and published reports on Co-ordinated Project Information (CPI). These offered a basis on which to build IT systems, but technology develops faster than people can organize themselves to make use of it.

The latest techniques for exchanging multimedia data are based on the World Wide Web, which also gives access to a vast body of unstructured data in which much valuable information is buried. New techniques are needed for mining this data, for providing a single entry point to the growing number of sources such as the proposed Construction Industry Gateway [45]. At present, when product data is wanted in electronic form, all those who can deliver it are rushing into new media, having their sales literature scanned, setting up Web sites and sending out CD-ROMs. They would welcome a structure into which each could fit, provided that it did not inhibit their promotion. Initiatives, such as the Industry Foundation Classes for object model definitions, and Barbour's Construction Expert, could provide suitable frameworks.

Those who are shrewd enough and have the resources to invest long term, are placing their money on good, timeless data. Bill Gates of Microsoft has been buying the reproduction rights to great master paintings, not only to display on the flat computer screens on the walls of his new house, but to profit from their eternal value. He was not quite so shrewd in assessing the significance of the Internet since, at one time, only two computers at Microsoft were allowed to be connected to it for security reasons. It was only the vast resources of the company and the free distribution of Microsoft's Web browser which allowed them to catch up with Netscape when they realized the future significance of worldwide communications [63].

Enjoy the vast resources of the Internet now while they are virtually free but do not expect to find really valuable data until access to it has been fully protected and charging mechanisms are tried and tested. Only now is good information starting to command its true value.

7.2 *Marketing is vital*

Throughout the 1980s the Construction Industry Computing Association evaluated many CAD systems, going into great detail to compare their capabilities for drawing, visualization, data management and ease of use. This culminated in an epic report by Paul Richens *Microcad Software Evaluated* [46]. Points were allotted for all these aspects of performance following tests on twelve of the most popular programs on PC and Macintosh computers. Conclusions were that Macintosh systems were generally the easiest to use, and that there were some systems particularly good at drawing production, while others specialized in 3D visualization. There was no best buy since CICA's experience was that users' needs vary widely. The object was to enable architects and engineers to match their needs to the capabilities of the most suitable system. However, the reader of the report could add up the points allocated from a series of tests and come to the conclusion that AutoCAD was a good, all-round system. It was not the best at any one thing but had made great strides since its launch as a primitive system on the first IBM PCs.

Autocad's rise to numerical domination was recorded by Brian Gott in his multi-client study, *The UK AutoCAD Market* [47] for Cambashi. In 1990 Autodesk had a chain of dealers who competed with each other but from whom Autodesk received only a 7% return. The dealers retained 44% of sales, and hardware suppliers retained a similar proportion. Some dealers were restricted in selling other products but they made more from selling this low cost system and their sales grew. Third party software developers were encouraged to add breadth to a basic drawing package and, by the mid-1990s, AutoCAD had started to dominate the construction market offering the full range of CAD functions with ambitions to serve the whole design and production processes in several industries. This growth could not continue for ever and margins had to be increased when Bentley Systems regained control of Microstation and provided stronger competition. There are still many CAD systems on the market and they all have excellent facilities, but it is their marketing strategies and strengths which maintain the success of a diminishing number of these.

Autodesk's revenues grew rather more slowly than those of some CAD vendors in the 1980s but market share is everything and the user base becomes a market for new products. Other computer firms could not afford to wait so long for their return. Both the political and business ethics of the 1980s were short term and concerned with making a fast profit.

If IT systems suppliers do not have patient shareholders or banks behind them, the lesson for the future should be to offer a safe, steady selling product to provide base income while they work on more speculative developments which might just change the world.

7.3 Succeeding technologies

The problem with introducing new information technologies is that the cycle of replacement is very fast. Long-term thinking is difficult when the next product which threatens yours is announced soon after yours has reached the market. The timing of IT announcements is an art designed to create panic among competitors while not stretching the imagination of customers too much. Entrepreneurs like Clive Sinclair, who was first to market with a digital watch, pocket calculator and a very cheap computer, are often overtaken by those who exploit the attention gained by the innovative products. British technology owes him a debt, but the lesson is that it does not pay to be a pioneer, unless you have the resources to update your first products very quickly. It is said that software only becomes reliable by its third release – look at Windows. Few ever heard of version 1. Version 2 disappointed, and it was Windows Version 3 which finally provided a user interface comparable with that of the Macintosh launched ten year earlier. Windows 95 maintained the improvement while demanding even greater computer resources, but the market requires regular releases and these often add unnecessary refinements.

Transmission of electronic data for construction has a long history. The first timesharing systems were applied to the ever present need for building product data. Accessing the Commodity File in the early 1970s, on a computer bureau via a typewriter terminal with a slow acoustic coupler, was painful and discouraged further trials until the 1980s when the BT Prestel service was tested by the National Building Agency on various types of information, including the software directories which I managed to compress into the restricted format. This was the revolutionary new videotex service based on TV screens with automatic dialling. It was still using low-speed lines and the display was restricted to some 20×40 pixels. This did not impress those in construction obsessed with presentation, although certain fast-changing data, such as the NEDO cost indices for inflation of building prices, was very suited to the medium. Videotex was little used in the UK, except in financial dealing rooms and by travel agents, but it was the basis for the successful Minitel service established, with much public expenditure, in France. This created a culture for on-line data access which is in advance of that in other parts of Europe. On-line databases are well established in fields other than construction, particularly in science where the existence of related work and published papers has to be thoroughly researched.

Few would have predicted that a general information service of global availability and low cost would attract the potentially vast business of construction data. There had been too many prototype communications systems promising so much and lasting such a short time. But, as is so often the case, a service from outside the industry became so attractive and widespread that it superseded most of the efforts from within construction. Fortunately digital data is transferable and, when the commercial opportunities of the Internet are fully established, construction data will grow rapidly on this universal medium. It needed the subsidy provided by the US military and by universities around the world to establish this network of networks, but it has the great merit of not being under the control of a single, exploitative organization.

The value of the Internet to construction would increase if the major information providers could agree between themselves to conform to an indexing structure which would allow search engines to hunt through the disorganized wealth of resources.

7.4 Software evolution

As each new technology is announced or emerges from the research phase, there are expectations that it will supersede what has gone before. Expert systems were hailed as a new, intelligent form of software which would revolutionize most applications by incorporating human experience, explaining their logical processes and giving probabilities for their results. Where did this technology go? There are few systems which go by this title now, but much software is the more intelligent as a result of knowledge based techniques.

The most successful expert system in construction was Elsie developed for the RICS at Salford University under Peter Brandon. This provided early cost advice on specific building types using the expertise specified by experienced surveyors. Particular clients, such as ICI, found they could explore the potential for a new project cheaply and quickly before committing themselves to commissioning a design. Such techniques have now been incorporated into other cost advice systems.

Figure 7.2 Network computer – the JavaStation by Sun with diskless processor.

The frontier for research has moved on to neural networks which learn as they process data. One application is in building management systems where commissioning of new service installations could be speeded up by self-teaching controls. The ultimate goal is to emulate the working of the human brain. In due course, when we understand more about how the brain works, we can expect neural networks to be absorbed within general programming techniques.

New information technologies do not often fail; they become assimilated within the general development of technology.

Marketing power blocs affect the choice between different technologies. At the time of writing the Network Computer is about to reach the market to counter the dominance of Microsoft and the overloaded PC. This is another manifestation of the cyclic phenomenon of centralization/distribution. All businesses go through waves of decentralization and outsourcing followed by gathering resources to exercise greater control again. The fashion of the early 1990s was for outsourcing computer services. Keen prices were often quoted for the first years of operation and, when the firm had lost the capability to run systems in-house, these prices often increased leading to disillusion and eventually a return to setting up an in-house department.

Recently, *Management Today* [48] featured a new business relationship being established between a courier service and a chip fabricator where speed and reliability of delivery was vital. 'It binds them to us as if they were our own distribution department' said the MD of the fabricator, claiming a new management breakthrough.

The Network Computer has a place in larger corporations where data can be managed centrally and the networks are reliable. It used to be called timesharing – for those who have short memories. For lone PC users who want to know that software and data are under their control, the complete processor with its own storage will be preferred, even if its management can be a nightmare. The specification of the Sun JavaStation network computer launched in 1997 was: 8–64 Mbytes of memory, with the Java operating system and HotJava Internet browser, fast communications, good colour screen, keyboard and mouse, but no disc drive (*Figure 7.2*). It will take a change of culture among most users not to give up the facility to back up their data locally or retain their own software. Managers of networks of such computers will regain control of the software and reduce the danger of users importing their own software and the viruses which can come with it.

A study of World Wide Web usage presented at Internet World in December 1996 [49] showed what percentage of Web users were making use of its main facilities:

Table 7.1 Percentage of World Wide Web users making use of its main facilities

Email	32%
Research	25%
News	22%
Entertainment	19%
Education	13%
Chat lines	8%

There was little commercial use of the Web, however the number of commercial sites grew from 23 000 at the end of 1995, to 220 000 by the end of 1996, so there is a potential which awaits user confidence in the security of on-line financial transactions.

The two camps competing to dominate the Web are Sun and Oracle with Java, and Microsoft's ActiveX. 5000 developers for Java were claimed at the end of 1996 but Microsoft could use its market dominance to establish ActiveX as another *de facto* standard. Other influential companies, such as IBM, had yet to declare their hand.

There is little doubt that the next fields in which battles for domination of information technology standards will be fought are the means of access to the World Wide Web and the high quality commercial data which it is starting to offer.

7.5 *Understanding the need*

The very earliest applications of computers in construction were by pioneers of computing from outside the industry. They assumed that automation of the design process was a definable task, that designers could agree on the steps they went through, and that these could be encapsulated in software. This proved to be a false hope and the earliest, successful applications tended to come from construction industry people who became enthusiastic about computers. Since then it has been more essential to understand the problem than to be an expert in computing as well. Another application in search of a use was network analysis, with critical path techniques used in the space race providing a sophistication which the available data in construction did not warrant. Simpler programs, with barcharts linked to resources, proved more generally successful but the more advanced critical path techniques offered in software have encouraged the planning of construction projects in greater detail with improved performance as a result.

The most successful applications are those which address routine tasks such as word processing, accounting and draughting. Justification of these is only a matter of demonstrating that the hardware and software can produce a better and faster response to the need than the established method. Most of the obvious applications have been addressed and it is new combinations of these, and radical changes to existing processes, which will support the applications of the future. Success is likely to lie in the gradual convergence of systems rather than expecting wholesale change. The paperless office is a goal which recedes into the future but projects, such as the Bluewater shopping complex near Dartford, are relying upon electronic communications with data formatted according to current standards and video of progress transmitted to the project offices and those of the client in Australia.

The virtual project team is a concept for the future with the participants collaborating from their remote offices, wherever in the world they may be. The reality is that advanced projects, such as the Glaxo Research Campus [5] or Terminal 5 at Heathrow airport, assemble the main parties in a single office to ensure there is a meeting of minds as well as integration of systems.

The concept, which will be hard to forsake, is that of the document. Building information is still seen as a series of traditional items: drawings, schedules, bills of quantities, orders and invoices. The idea that these could be part of a continuum of data of all types, combined in different ways for different purposes, is a difficult one and, although we should stop thinking of IT as a series of discrete processes, it is useful to retain the concept of the document for the foreseeable future.

Computer systems are seen currently as discrete aids to specific processes but, with communications now sufficiently powerful, and with a growing base of electronic data, they will soon be seen as part of an overall process – the transfer of data from the mind of the client, through the computers of the project team, to the control and management of the building.

7.6 Costs and benefits

The big question, which any business will ask about a new technology, is what it will cost and when the benefits will exceed the cost. The first is relatively easy to answer, although timescales for its integration into business processes and costs of training can be hard to estimate. Benefits can be fully quantified only if the application addresses a current function and the cost of this is known, but even then only with difficulty. The cost of a word processing system can be measured over time against a reduction in secretarial staff, less time spent retyping documents, and reuse of existing documents. But what about the value of better quality presentation and the tendency for more revisions to be required because they can be done more easily?

If word processing is as complex as that to quantify, what about CAD which appears to replace a manual draughting function but is capable of doing much more in a very different way. Early attempts to justify CAD were inconclusive without running the same project both manually and using CAD, side by side. Draughting productivity factors were quoted ranging from 10:1 for drawing grids or repetitive work, to ratios in favour of manual work for initial sketch design. Those who believed in the technology could choose the ratio that suited their case but there was also the benefit of being seen as a technology pioneer. Did this really balance the cost of early CAD systems at £100 000 per seat?

A study of the benefits of IT by CIRIA in 1995 [50] was based on case studies of contractors and civil engineering consultants. The former produced very little in the way of cost benefit studies, while the latter generally said they had to have the technology to obtain the commission and that applications, such as structural analysis, were essential to their work. The exercise of balancing cost against benefit, which still had to be done to obtain funding, depended upon the intangible benefits: a better service to the client, a more co-ordinated project or fewer errors on site. The earlier in a project that potential problems can be spotted, the less expensive it will be to solve them. Watch out for construction simulation software to anticipate such problems, and programs for emulating safety situations for the CDM regulations. These may be hard to justify unless there are extra fees for fulfilling these roles and the cost may have to be borne by the client. Aware clients will expect such systems to be part of the normal service but their consultants and contractors will have to reassure them that they have the right systems in place.

The most critical factor with IT investment is time. Can the benefits of a new technology be realized before the technology is obsolete or the systems need replacing? With expensive hardware that was always a problem – software can be updated and a responsible supplier will keep up with new developments. Data should have a longer life and the problem of transferring it to new systems, where there are standards in place, is becoming less. A simple cost/benefit model and related graph (*Figure 7.3*) allow variation of the cost of the initial investment and the period over which this is recovered. The subsequent period of net benefit will last as long as the technology is current. It becomes more complex when interest rates, replacement of hardware and maintenance of software are taken into account but, providing the application is still useful, good software systems should be adapted to the latest technology.

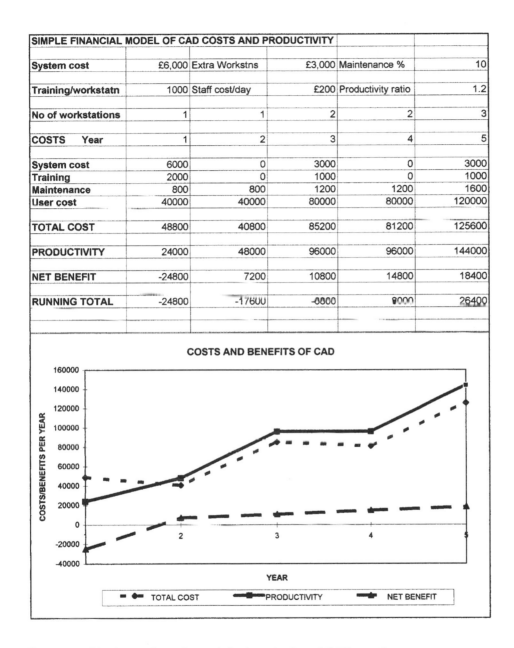

SIMPLE FINANCIAL MODEL OF CAD COSTS AND PRODUCTIVITY					
System cost	£6,000	Extra Workstns	£3,000	Maintenance %	10
Training/workstatn	1000	Staff cost/day	£200	Productivity ratio	1.2
No of workstations	1	1	2	2	3
COSTS Year	1	2	3	4	5
System cost	6000	0	3000	0	3000
Training	2000	0	1000	0	1000
Maintenance	800	800	1200	1200	1600
User cost	40000	40000	80000	80000	120000
TOTAL COST	48800	40800	85200	81200	125600
PRODUCTIVITY	24000	48000	96000	96000	144000
NET BENEFIT	-24800	7200	10800	14800	18400
RUNNING TOTAL	-24800	-17600	-6800	8000	26400

Figure 7.3 Simple cost/benefit graph for introduction of CAD over 5 years.

Value added benefits to the user's business should not be forgotten even if they are hard to express in accounting terms. In future there should be better measures for valuing intellectual property rights and the ability to have the business processes of a company or standard design procedures built in to the corporate information system.

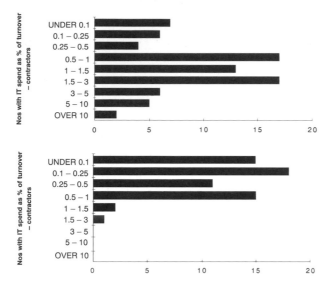

Figure 7.4 Range of percentage spend on IT by large consultants and contractors. *Building on IT for Quality*. CICA, KPMG. 1993.

7.7 Market characteristics

For someone looking at construction from the IT supplier's point of view, the market may seem homogeneous but it is in reality vast and various. There are some 200 000 companies in UK construction of which 150 000 employ fewer than 10 people. Consultants form only about 12 000 of these companies while the figures for contractors have been distorted by the large number of self-employed tradespeople.

Levels of spending by the top 800 firms in 1993 [33] showed a median spend for contractors of 0.25% of turnover while consultants spent about 1.5% of their fee income (*Figure 7.4*). This includes hardware, software, communications and dedicated computer staff. With a £50 billion workload projected for 1997, this implies at least £250 million p.a. spent on IT by contractors and £75 million p.a. by consultants, assuming total fees represent about 10% of turnover. Building materials producers, who are generally classified as manufacturing industry, form another large market for IT products.

Even within these three major subdivisions of construction: the consultants, contractors and materials producers, there are wide variations. For contractors these are primarily between general contractors and specialist subcontractors. There are also civil engineering contractors and housebuilders. The types of systems appropriate to each of these groups depend upon their specialism, for example building services subcontractors do much design work.

The types of contract currently in use include: design and build, management contracting or the traditional tendering against drawings and quantities. For consultants there are the main groups of architects, engineers and surveyors each of which breaks down to specialisms such as civil, structural and building services engineers, and quantity, building, land and property surveyors. There are also new professional groups being formed to meet new responsibilities or legislation: project managers, CDM supervisors and adjudicators.

Computer applications for these groups are similar as regards standard office systems, but diverge in the accounting function and are widely diverse in the specialist technical areas. For instance CAD is much more widely used by housebuilders than by general contractors since they carry out their own design work and use it to sell their houses. Civil engineering contractors use project management much more since they are working on larger projects. Among consultants structural and services design software is confined to specialist engineers while bills of quantities programs are largely used by quantity surveyors. Product manufacturers combine manufacturing and stock delivery systems with CAD systems for installation drawings, and communications for regular trading relationships with builders merchants and large customers.

For those selling IT systems to the construction industry these variations should be recognized and the systems tailored to their needs. More of the larger firms are now working internationally and there are major differences in the accounting practices and specialist engineering codes followed in different parts of the world. The US has its own methods which influence international contracts, but UK methods are widely used in Commonwealth countries and the Far East. Euro codes for structures are now defined but not yet widely used, while those for services calculations are making slower progress.

Construction represents a vast market but its complexity need to be understood by those developing software and supplying its systems.

7.8 Standards and promotion

In a field where there is a great diversity of applications software there is a tendency to select systems which conform to standards or come from well known suppliers. CICA listed some 1500 applications for construction in 1996 [44]. The standards include the type of computer on which they run and the operating system, with the Intel based PC and Windows as the most popular combination, and the relevant codes on which the application is based. The latter are particularly important for engineering software where the legal liabilities connected with its use are of particular significance. The prominence of the supplier is dependent upon market share or their notoriety as an up and coming supplier of future *de facto* standards.

Most standards are *de facto* and result from market dominance so that standards and promotion are intimately coupled. The secret of success is to make a general purpose IT product available free, or at low cost, so that it is widely adopted and is incorporated into larger systems. Examples are the MSDOS operating system, the HTML language on the Internet and the Netscape browser for the World Wide Web. None of these are formal standards but are widely used and any successors tend to have to incorporate them. For instance Windows was originally based upon MSDOS and Windows 95, even though independent of MSDOS, still runs MSDOS programs. Formal standards have generally not kept pace with the rate of development of the technology but contribute by endorsing particular approaches. System suppliers and their user groups provide much useful input to standards in the hope that they will reflect their particular technology.

Promotion is aimed at making a product into an established name in a particular market whether through market share or getting noticed. The IT industry is notorious for rumours of the next wonder product and announcements are made, often more in hope than in expectation. It is said that hopeful companies will deliberately infringe patents to gain publicity from being sued by a well known name. An example of hopeful, and probably effective, promotion was Larry Ellison of Oracle announcing the Network Computer during 1995 and continuing to press its claims as the alternative to the overloaded PC. The need for some way of balancing the domination of Microsoft found ready supporters and a number of companies, including Microsoft themselves, announced Network Computers for delivery during 1997.

Success in construction depends upon securing some prominent users who will pub-lish case studies of their success with the new product and encourage their trading partners to adopt the same systems. Partnering is a new approach in an industry whose members have traditionally worked with a different team for each project. Longer term relationships will allow systems and communications to be set up for more than the 1–2 year life of a building contract. This period hardly allows systems to be tested fully let alone pay for themselves. Partnering involves the client also and, if the client believes that better systems will result in better buildings, resources and contracts will be adjusted to provide these. The leading firms introducing partnering are BAA, with its Framework Partners who bid for a five-year relationship, the retailers and the banks.

The effect of leading clients on the technology of consultants and contractors is considerable and, for an industry with many parts which has often failed to co-ordinate itself, the influence of client industries more committed to IT can only be beneficial.

7.9 Summary

From the factors which have affected the success of particular IT systems in construction and some more general considerations, it is possible to set down some guidelines for assessing future technologies. *Timeliness* is important since a product introduced before either the technology it is based upon, or the market, is ready can lead to succeeding products reaping the benefits. *Marketing* will provide intelligence on these essential factors and effective promotion will create demand, something which is very necessary with high technology. *Evolution* can be rapid and those introducing new systems must be prepared to revise them regularly but resist users' demands for extra features if they are likely to make systems less efficient. Providing products which meet the particular needs of an industry like construction may require *specific solutions* and these can only be developed in conjunction with that industry. If they meet a need then users have only to estimate the value to their business and the period within which they can expect *benefits to exceed costs*. Even if this is not convincing in accounting terms, the value added benefits may be considerably greater but will need to be quantified in some form.

The construction market is large, even if it consists of many parts. It should use its muscle to set standards for itself and to influence IT suppliers. It also needs to co-ordinate IT use between its parts and its poor record in doing this until now should change to meet the needs of high technology clients.

8

Some promising developments for the future

From the analyses in Chapters 5, 6 and 7, the most influential factors and their probable effects on existing systems, can be applied to some promising new developments. This is not with the aim of predicting the future but to offer some methodology for assessing forthcoming technology and to test it on recent developments which are topical at the time of writing. There are many new products in research at present, and there will continue to be many new ideas, which are presented as original solutions to fundamental problems. Most of these evolve from existing technology and few will address the particular needs of construction, but those involved in building design and construction need guidance to assess their significance and to decide how to apply them most effectively.

8.1 General technology development

In Chapter 5 the following patterns emerged from the study of general technologies:

Computer hardware systems have changed from being offered in a range of sizes, to physically small systems connected to local and global networks. Processing is currently distributed to the peripheral PCs in peer-to-peer networks, but there is a move to centralize it again using client/server networks which store software and data centrally and supply them to network computers with no local data storage.

Operating systems have reduced in number as computer precision has evolved from 8, to 16, to 32 bit processors, with 64 bit processors following. The market is dominated by Microsoft Windows, with niche systems like UNIX holding their own, and Web based languages, such as Java and ActiveX, making rapid progress.

Standards going through formal channels are too slow for information technology and are most effectively delivered in partnership with industry *de facto* standards.

8.2 *The family of S-curves*

The various paths on the S-curve graph (*Figure 8.1*) can be illustrated by examples of past technologies, many of them now forgotten, but they will be followed by many similar examples.

A. *Fails quickly* Examples [10]:
- Apple Lisa overtaken by the Macintosh. 2700 were buried in a Utah landfill.
- The DEC Rainbow was not IBM compatible and could not format its own disks.
- Microsoft Windows Version 1 – a pale shadow of the eventually successful Windows Version 3.

B. *Declines after being overtaken* Examples:
- Visicalc, the original spreadsheet, superseded by Supercalc, Excel and Lotus.
- Clive Sinclair's various computers which helped stimulate sales of the more advanced Commodore Pet and Apple II.

C. *Eventual decline*
This happens to all technology, even MSDOS which is technically obsolete, but all new office systems still have to support it to maintain legacy data and software.

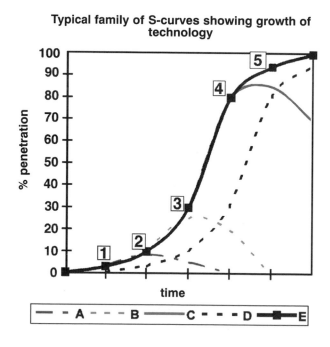

Figure 8.1 Typical family of S-curves showing growth of technology. For key see *Figure 5.6*.

D. Following technology
Represented currently by Windows NT which seems likely to supersede Windows 95 and even UNIX, but may be overtaken by Web based operating systems.

E. Ideal growth
Represented by the S-curve itself which typically applies to communicating technology where, once a certain proportion of the market have it, the rest are obliged to follow. Examples: Telephone, Fax, Electronic mail.

8.3 *Summary of factors affecting general applications*

In Chapter 6 the deductions made from factors affecting general applications were:

Wordprocessing – a basic requirement which awaited ideal hardware and software and will only be radically changed by adoption of speech input and the 'less-paper' office. There is a move towards more data, such as standard letters, being kept in databases. There will be substitution by electronic and voice mail and more reuse of existing documents, but the basic need, to process words, remains.

Spreadsheets – a new application which created a general need. New factors might be portability, integration, predefined financial models for particular functions, and more graphical output. There will be more integration between office systems through linking and embedding spreadsheets within other documents.

The Internet – an unspecified, but generally desired, capability to transfer data anywhere. It grew quietly in the academic community with public funds, and became a phenomenon when it was discovered by the business community who now need to make it more secure and commercial. *De facto* standards were accepted and there is now exponential growth with eventual links to many homes and every business.

Computer-aided Design – had a slow start since hardware and software were very inadequate initially, but systems eventually met a general need for drawing production and have added more sophisticated presentation and modelling features in recent years. Suppliers have moved up-market into Geographical Information Systems and complete design support systems which provide data for use within Facilities Management systems.

Communications – on-line access has now caught up with CD-ROM usage thanks largely to the Internet. Hybrid systems offer the best of both these technologies. Email has taken off in construction with 15% of large firms in construction using it in 1992, growing to 40% by 1995. ISDN provides the higher bandwidth needed for graphical data and ATM may provide the longer term solution to sending large volumes, particularly when optical fibre is installed to the desk and home. EDI will eventually be delivered by the Internet and used locally on Intranets.

8.4 *Questions to be asked about future technologies*

When assessing new developments the following should be considered:

Are there any more general needs which IT has yet to address? Some have been identified but are not yet viable through lack of promotion or essential data. Others may result from the need to protect and value Intellectual Property Rights more effectively, so that transfer and publishing of designs will be more secure and asset values can reflect investment in valuable data.

What further convergence could there be of existing systems? All related applications could eventually be integrated but they also need to be capable of standing alone. At which stages of a continuous process will human intervention be desirable or necessary? Ideally there will be tools to monitor progress and detect problems so that people remain in control.

What hardware and software developments will open up new applications? Continuing improvements in cheapness, capacity and speed of computers, plus communications and intelligence, will make existing applications usable by all, but there will always be some who are resistant and their views should be respected. Portable computing and self-teaching software will offer new opportunities.

How will the market change? When will current market leaders be overtaken? It is only necessary to look at the changes in IBM to see that size does not guarantee continuing success. In fact, in a fast changing field, manoeuvrability is more important. The construction industry is changing. It can specify its needs for IT better and these will be served best by small groups of collaborating suppliers.

Will the cost benefit ratio continue to improve? Costs of hardware have been cut below levels anyone thought possible but must stabilize at some time. The rate of increase will remain below that of labour and travel. It is hard to see IT ceasing to be competitive and it will compete in a growing number of areas of business.

From where will future standards come? Formal standards have not been keeping up with the demands of IT except in the field of communications, and will need to be developed more closely with, if not by, business. Formal endorsement across trading groups such as the EU or, preferably, worldwide, will still be necessary.

8.5 *Factors affecting construction applications*

In Chapter 7 the deductions which can be made from the factors affecting construction applications were:

Computer-aided design – awareness of its benefits is now widespread and all large design firms have some investment in systems and expertise, but there will be continued growth in numbers of workstations and in the use of CAD by contractors. Project data management based on CAD models, and new techniques for presenting these, will ensure continuing interest and development.

Accounts – are a basic need for all businesses, more essential to contractors, but of growing importance to consultants. There will be more integration of specialist modules and greater use of resource control, with projections of future cash flow using simulation based on analysis of past projects.

Estimating – there have always been a number of different systems available and much marketing of these, but growth in the 1990s depended upon the rise of less conservative estimators. Quoting the right price needs good purchasing systems and is still an art combining experienced judgement with good data.

Bills of quantities – growth has been slow since the first programs became established and there may be a decline if simpler forms of bills are used than those based on the Standard Method of Measurement. Contractors will still need quantities in some form although these will come eventually from CAD models.

Project management – the flat growth curve (see Figure 6.10) shows that this is a specialist application although it has been available for some time. Its use is now essential on large projects and portable computers allow networks to be updated on site by managers rather than by depending on site staff to keep them continually up to date. Networks for all projects will be linked to assess total resource requirements.

Structures and services – analysis software is linked to published methods which structural and services consultants are obliged to use. It is therefore an essential application which needs a new version each time codes are revised. Integration with CAD is a goal which is yet to be fully achieved.

8.6 *Projections of some recent technologies*

Network computers

At the time of writing these are largely a marketing phenomenon with some support for the idea from users who have problems managing overloaded PCs. Future success is more likely with the large, corporate user supported by an efficient network, and the latest software maintained centrally. Like most users frustrated with timesharing from dumb terminals in the past, they are likely to smuggle their favourite software onto their portable machines. The small firm, or lone computer user, will see little benefit from dependency on data and software from networks although occasionally used programs will be downloaded from time to time. Such software might include energy analysis, specialist structures and exchange software needed for particular projects. *The success of network computers will depend on whether the managers of large networks regain control from the users.*

Java v ActiveX

Internet based languages will become more important and the latest offerings are Java from Sun, and ActiveX from Microsoft. Marketing is the key to their success and this is linked to the use of the World Wide Web and network computers.

> According to Dick Pountain [51] the Object Management Group settled on the OpenDoc standard in 1996. This conflicts with Microsoft's ActiveX. Sun's Java can talk across the Internet and is backed by Netscape's 38 million copies of Navigator, which is ready to accept Java-based plug-ins. This seems to have panicked Microsoft into a compromise: it will hand over its ActiveX object technology to a customer-driven, open standards body. So, he concludes, customer pressure might finally force a convergence of object standards.

Microsoft is a formidable influence against a return to centralized computer networks but this is a cyclic phenomenon and will undoubtedly return. There will be other languages and browsers and the future is unclear apart from the likelihood of their incorporating facilities such as Applets, for fast publishing of data on the Web, and knowledge agents for helping to find data matching a user's needs.

The growth of languages like Java and ActiveX is likely to be rapid.

Intranets

Succeeding technologies often creep up on established ones without the world noticing and, when they are established, seem an obvious substitution. After the dominance of Novell's proprietary networks, which were an advance on previous networking protocols, it now seems obvious that the TCP/IP conventions used on the Internet should also be used for internal networks or Intranets. They allow data to be published internally and externally in the same form, since many firms, large contractors in particular, are afraid of providing full Internet access to all staff with the possibility of time being wasted and viruses corrupting their internal networks. The solution is a self-contained Intranet on which useful internal data can be held and from which data for the attention of the outside world can be published. Wireless LANs are likely to replace some of the cable-based ones provided that a sufficient range of frequencies is available, and these will extend networking to organizations which cannot accommodate additional wiring.

It seems inevitable that the same type of networking software will be used for internal networks, or Intranets, as is being used for the global Internet.

Interoperability and the virtual project team

General technologies will continue to change in spite of the specific needs of construction. There are significant niche applications which construction develops to suit its own needs. The Industry Foundation Classes is one such, applying interoperable objects to the elements of a building: from initial concepts, through specific components, to the provision of lifetime data on building operation. Such a body of data, developed through international collaboration by the International Alliance for Interoperability, is essential for building design to take the next step in realizing the benefits of CAD. Draughting and visualization is widely established and 3D models, shared by all parties and including non-graphic attributes, are the basis for the next leap forward. These will enable a project team to work from remote locations and will eventually lead to the virtual project team. The communications technology is in place, and the awareness of leading firms and their clients is encouraging them to commit resources. Research projects, such as BRICC [52], have demonstrated the possibilities. What is now needed is a body of standard object data and the will to implement this on major projects.

CAD vendors are supportive and, provided this data is defined soon enough, they will implement these objects to make the most of their system's capabilities while maintaining the interoperability of these with those on other systems.

Formal and *de facto* standards

The process of defining formal standards, published by BSI, CEN and ISO, seems to be out of touch with the fast changing needs of information technology. In construction there is a good basis of British Standards, many of which have been adopted internationally. Many ISO standards, except ones like ISO 9000 for quality assurance and ISO 10303 STEP, are in small sections, covering the only areas in which agreement could be obtained, and are little used.

Meanwhile *de facto* standards, defined by market leaders or small groups of businesses, dominate IT. A relationship between these and the formal bodies is developing and the pattern adopted by ANSI in the US allows pressure groups to develop their proposals into formal standards after due public comment. In the UK commercial proposals are kept at arm's length and the tendency has been, as in the CAD layer standards, for commercial interests or their user groups, to take a published standard, develop it further to suit their particular technology, and give it the promotion that standards organizations can rarely offer. This can be a very fruitful way of proceeding provided it is clear which is the official standard and what has been developed from it.

Formal standards are successful, indeed essential, in some areas such as telecoms with those from CCITT, but others, like Open Systems Interconnection, have merely proved to be an idealist influence in the background, rather than taking the technology forward. *De facto* standards will continue to prevail and they should be developed using the methodology established in the USA. The European Union could follow this route since it will play a larger role in future, superseding national standards in its member countries.

National governments will still need to provide support for representing their interests and to ensure that commercially led standards, developed internationally, meet a common need.

Voice recognition and speech technology

The ability of computers to generate and recognize sound has existed for about 20 years. Why has this technology not replaced typing and done away with the keyboard which, together with the cathode ray tube, keeps computers larger and more cumbersome than they need be? Delivery of sound is now firmly established with the introduction of the multimedia computer, but this does not present the same problems as analysing continuous speech. Here systems trained to recognize individual commands have been effective since the 1980s, but deciphering continuous speech is still a big problem. However, the number of words per minute, which can be recognized and stored as characters, is growing steadily.

The problem lies with the complexity of language and not having an ASCII keyboard equivalent for our voices. The goal which will change international communications is simultaneous translation. One of the telecom companies announced a service in 1994 which would accept English at one end and output Japanese at the other, or vice versa. It was still not widely used by 1997. This is another area in which the solution lies with a hybrid use of systems and people. Voice recognition systems still need to be trained for a particular voice, but neural networks could help them to learn more quickly. The resulting transcript will contain obvious mistakes which will have to be edited. There will also be more subtle errors which could cause misunderstanding, or offence, in international business [51].

It is predicted that speech recognition will be a standard feature of PCs by 2001.

The less paper office

This has been a goal since words first appeared on screens. Why print when most readers have access to a screen and electronic communications are so much faster than conventional mail? There are two main reasons: many do not readily believe that electronic data is permanent and like to have an overview and to browse through documents. There is also the legal profession who prefer to see words on paper with a date and a signature, and electronic commerce is not yet quite so clear in this respect. Yet we do allow vast sums of money to be transferred electronically and our insurance and investment records are held in this form whether or not this is fully trusted. So it should only be a matter of time before other types of electronic data become accepted.

For those who grow up with electronic commerce, there will be greater trust in electronic data. The growth of telephone banking is one manifestation of this. Document management systems compact vast volumes of paper files and archived drawings. They can record dates of revision, to whom documents were issued, and can be accessed from anywhere. Multimedia systems provide better browsing facilities than any paper publication. They offer instant indexing and the facility to see a summary then go into more detail. A well-designed system, such as the CD-ROMs published by Microsoft and Dorling Kindersley, or the Building IT 2005 report, provides a map showing the content of the disc and the means of navigation *(Figure 8.2)*.

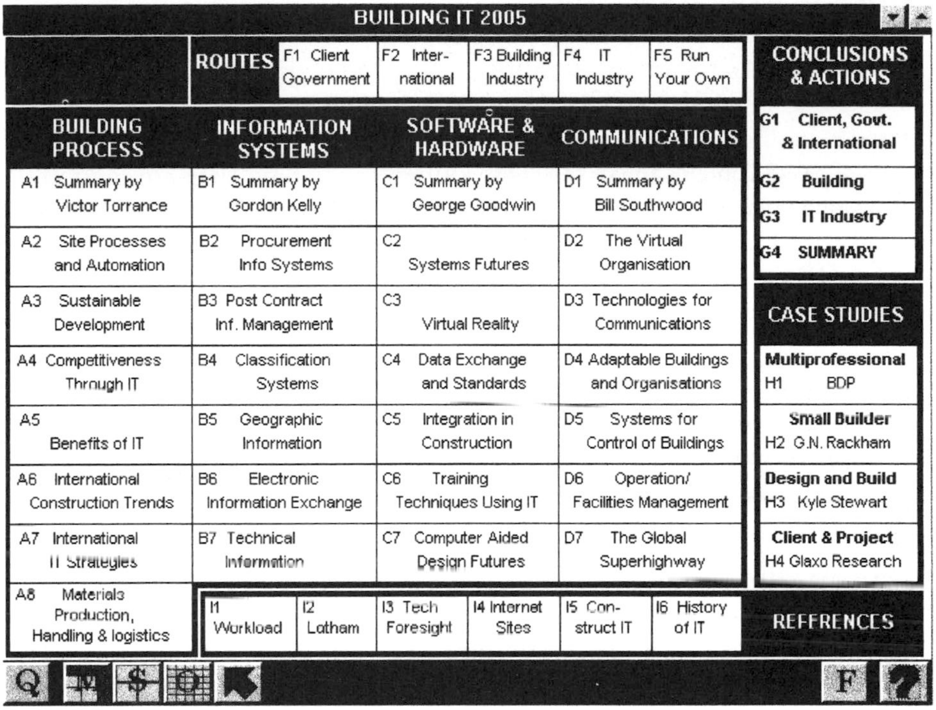

BUILDING IT 2005						
ROUTES	F1 Client Government	F2 Inter-national	F3 Building Industry	F4 IT Industry	F5 Run Your Own	**CONCLUSIONS & ACTIONS**
BUILDING PROCESS	**INFORMATION SYSTEMS**	**SOFTWARE & HARDWARE**	**COMMUNICATIONS**			**G1 Client, Govt. & International**
A1 Summary by Victor Torrance	B1 Summary by Gordon Kelly	C1 Summary by George Goodwin	D1 Summary by Bill Southwood			**G2 Building**
						G3 IT Industry
A2 Site Processes and Automation	B2 Procurement Info Systems	C2 Systems Futures	D2 The Virtual Organisation			**G4 SUMMARY**
A3 Sustainable Development	B3 Post Contract Inf. Management	C3 Virtual Reality	D3 Technologies for Communications			**CASE STUDIES**
A4 Competitiveness Through IT	B4 Classification Systems	C4 Data Exchange and Standards	D4 Adaptable Buildings and Organisations			**Multiprofessional** H1 BDP
A5 Benefits of IT	B5 Geographic Information	C5 Integration in Construction	D5 Systems for Control of Buildings			**Small Builder** H2 G.N. Rackham
A6 International Construction Trends	B6 Electronic Information Exchange	C6 Training Techniques Using IT	D6 Operation/ Facilities Management			**Design and Build** H3 Kyle Stewart
A7 International IT Strategies	B7 Technical Information	C7 Computer Aided Design Futures	D7 The Global Superhighway			**Client & Project** H4 Glaxo Research
A8 Materials Production, Handling & logistics	I1 Workload	I2 Latham	I3 Tech Foresight	I4 Internet Sites	I5 Con-struct IT	I6 History of IT	**REFERENCES**

Figure 8.2 Map showing content of multimedia. Building IT 2005 produced for the Construction IT Forum by the author working with Cambridge Multimedia.

The UK Technology Foresight report on IT and Electronics [4] forecast that 75% of paper would be replaced by electronic imaging and EDI between 2004 and 2009, and that the rate of creation of multimedia documents will exceed that of single media by 2009, although there were a significant number of respondents who believed this would never happen.

The concept of the document is unlikely to be given up by the construction industry yet, even if it is just a selection of data from a database.

Effects of IT on travel

This is another change to the workplace on which IT should already be having a major effect. The technology can deliver information electronically whether it is text, sound, image or video. However, rather like our reluctance to reduce our reliance on paper, there are elements of human contact which computers cannot, and may never be able to, simulate. Technology Foresight for the construction industry rated 'travel to work distances are halved by development strategies and IT' as the most significant benefit to the quality of life, and most of those polled believed that this would be achieved between 2004 and 2009. It was also highly rated for wealth creation but few office staff want to cut themselves off from the buzz and complex interactions of the office environment.

This book is written from a house on the edge of Cambridge, a city with traffic problems worse than most of its size. As I read my electronic mail in the morning, I see, across the fields, a line of cars building up on a country lane which is itself an alternative to a main road on which there is another line of slow moving cars, three miles from the centre of the city. The occupants of these vehicles make the same journey every day. They are comfortable and know how long it will take them to get to work even if it is a little more each year. Few of them would wish to work from home but there will be an increasing number of telecottages, or shared local business premises, for those working remotely from their employers or customers (*Figure 8.3*).

Another hybrid solution is for small businesses to form groups using electronic networks or shared office space in, or near, the homes of their employees. Those who are employed by large companies could work from home for several days each week and share office accommodation in the manner of the 'Space net' concept which the Chadwick Group is applying to the offices of Andersen Consulting. Office space is allocated to consultants only when they need it, and their possessions are stored on trolleys in lockers. The 'Space net' principle also links their various European offices as though they were one single office. Andersen Consulting reduced its space requirement in Paris from 10 000 sq m to 7 000 sq m in a more prestigious city centre location [53].

It will take a long time to convert most businesses to the idea of dispersing their workforce or sharing their office space, but the technology already exists to do this.

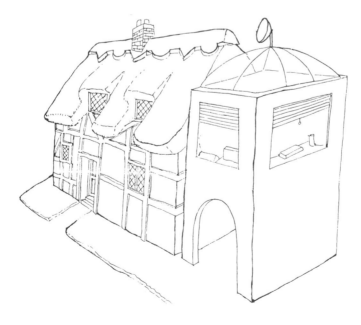

Figure 8.3 Image of the telecottage from *150 years of building*.

8.7 *Database technology and integration*

Behind every major software system is a database. This potentially vast collector of data in records and fields can be organized in tabular form, relationally or in object-oriented form. Object database records inherit data from each other and each record contains its own relationships with other records. This is another technology with a surprisingly long history; the basic concepts of the object-oriented approach were introduced in the Simula programming language in Norway during the late 1960s. Objects are meant to represent physical things – building components for example. They pass messages to each other and are grouped into classes. Messages are simply the name of an object and the name of a method which that object knows how to execute. For example a door would know that it requires a hole in a wall with a lintel over it. It would belong to a class of doors which would all know that they have a height and width and one or two opening panels. Each instance has specific values for these and a door schedule can be produced from such a database [54].

Object-oriented CAD systems depend upon data records in which geometry and appearance are only two types of field. Data on cost, performance, suppliers and maintenance can also be attached. It is then possible to add cost planning and quantities routines, to analyse performance and generate maintenance schedules. Unfortunately the data is rarely available until late in the project design phase and then may never become part of the project model. Clients for future projects, aware of the benefits of an interoperable model, will appoint consultants, contractors and even suppliers, at an early stage and invest in building a model before the design has developed beyond the stage at which this would be of greatest value.

The STEP standard for product models is one of the slow but promising standards being developed under ISO. Its construction Application Protocols will emerge as models defined in the Express language, during the late 1990s but events, driven by CAD vendors, will probably deliver the technology sooner than that. One ambitious initiative for the construction industry, which should define a wide range of building objects by 1998, is the Industry Foundation Classes. These are being defined by members of the International Alliance for Interoperability and will produce public domain definitions of widely used building elements, as Express-G models presented graphically for easier checking by non computer specialists (*Figure 8.4*). They will then be implemented by the leading CAD vendors involved. The EC SABLE project includes Autodesk Publishing, the French CSTB, Wise and Loveys in the UK, and the Irish Building Information Centre, and aims to produce a European set of building components distributed on CD-ROM and the Internet.

Integration depends upon access to common data. The future value of good libraries of building objects will be enormous, particularly if a framework is established and product manufacturers can be persuaded to conform to this. Classification systems, such as Uniclass, being proposed to bring together the existing systems CI/SfB and

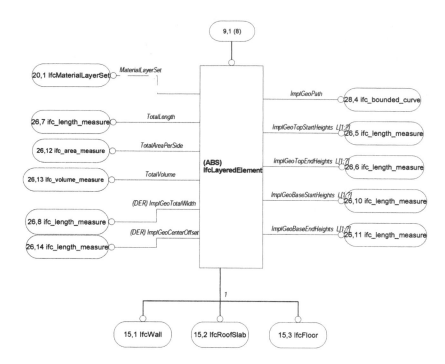

Figure 8.4 Example of an Express-G model for representation of construction objects.

Common Arrangement following the ISO Technical Report 14177 [55], are one aspect of such a framework. Product databases, such as Barbour's Construction Expert which I suggested would benefit from being linked to AutoCAD menus and is now accessible from AEC 5, could provide another such framework to encourage product suppliers, and the IFCs could add functionality.

All these efforts play a part but it will take a miracle for all those who would benefit from the technology to pull in the same direction for long enough to establish a common building model. Marketing of different products and CAD systems, and the wish of architects to design in novel ways, are likely to result in a continuing variety of ever changing approaches to the representation of buildings in IT systems.

8.8 Internet commerce

As with the provision of data generally, that on the Internet has initially been free. Resulting from this, what we find is mainly what the authors wish us to see. The spirit of the pioneers was for free exchange of ideas and information and it is an excellent medium for distributing research results. Commercial interests are starting to take a different approach. If sites are popular they can carry advertising, even if this has not yet achieved measurable success. World Wide Web sites can be made to record the number of visits to them and advertisements can divert users to the advertiser's own site, so it is an ideal medium and will surely come into its own. Commercial sites grew tenfold between 1995 and 1996, and these include advertising, paid for access to authorized users, and shopping.

Table 8.1 from Forrester Research [56] gives forecasts made in May 1996 for the growth of shopping sectors in $ millions.

Table 8.1 Forecasts for the growth of shopping sectors on the Internet ($ millions)

Segment	1966	1997	1998	1999	2000
Computer products	140	323	701	1228	2105
Travel	126	276	572	961	1579
Entertainment	85	194	420	733	1250
Apparel	46	89	163	234	322
Gifts/flowers	45	103	222	386	658
Food/drink	39	78	149	227	336
Other	37	5	144	221	329
Total	518	1138	2371	3990	6579

Devices for giving Internet users access to valuable information against a prepayment are called 'cookies'. These are small pieces of computer code loaded onto the user's computer which carry the right of access to the information to which a user has subscribed. There are some reservations because they can also be used to monitor the user's actions on the site. The following views are from a survey by *Byte* magazine [51]: this tracking capability makes users nervous. 'Commercial sites have every right to monitor your actions within the spectrum of their site', says one respondent. 'But they should not have the right to distribute that information to others.' Another says: 'Use of tracking data aggregated to eliminate individual identity is OK, but using individual's data is too great an invasion.'

Security is the key for getting valuable information made available. Concerns have been expressed about making credit card data available on the Internet. This should be of no greater concern than quoting card details over the telephone, although such numbers could be collected from the net on a much larger scale and by people in another part of the world. There are a number of services which claim to provide the necessary security such as: CyberCash, DigiCash, First Network Bank, First Virtual, NetBill and NetCheque. The major banks also provide services and these include MasterCard, Mondex, Visa cash and Barclays Netlink. Shopping services include shopping malls where a number of commercial sites are accessible from one point: BarclaySquare in the UK includes Blackwells bookshop, wine from supermarkets and the Virtual Vineyard, flowers, airlines, Eurostar booking and an Innovations catalogue where goods can be inspected and ordered.

A US Government report on electronic commerce [57], which includes commercial data exchange on closed networks, suggests the following principles:

* The private sector should lead – although the US government initiated the Internet.
* Governments should avoid undue restrictions on electronic commerce.
* Where government involvement is needed, its aim should be to support and enforce a predictable, minimalist, consistent and simple legal environment for commerce.
* Governments should recognize the unique qualities of the Internet.
* Electronic commerce on the Internet should be facilitated internationally.

The big question is whether to tax this form of commerce. The US government paper advocates that the World Trade Organization declares the Internet a duty-free environment. Only existing taxes should be imposed on such commerce with no new tax, such as the 'bit tax', which has been proposed. Are profits yet being made on the Internet which will attract the attention of taxation authorities? An Activmedia survey from September 1996 [58] quotes Auto-byTel car sales in the US as having made a profit on $6.5 million sales. Netscape sells $1.5 million of its software products by this means and the ONSALE auction house has a turnover of $45 million. Some 2.5 million people, mostly in the US, had purchased products over the Web by 1996 according to Forrester Research. **Other predictions suggest that, by 2005, 20% of household expenditure will be made over the Internet.**

This is more than the proportion of households having a computer at present and will depend upon the introduction of set-top boxes for televisions connected to cable services. These will turn the digital televisions most of us will have by the end of the century, into Internet devices with pay-per-view television as the incentive to buy them, and they will become the main device for electronic commerce and on-line shopping. These set-top boxes are already being developed and are expected to sell at $200–300. They may also be developed for general communications functions with software such as operating systems and Java.

Already six types of set-top box have been specified and standards will have to move fast if growth is to be controlled and set-top boxes are to intercommunicate.

Consumers may also wish to select their next home through these channels and the possibilities for displaying buildings on video or using virtual reality are exciting. Business will also use this medium for ordering products and transmitting commercial messages. With the arrival of the telecom providers as Internet service providers as well, it will be possible for them to include royalties for looking at high value pages, which can be priced at a few pence each, on their regular service bills [63]. This will avoid the problem of credit card transactions for each site visited and, for popular pages, many accesses at a few pence can generate valuable business. For projects based around computer networks a higher level of security is needed.

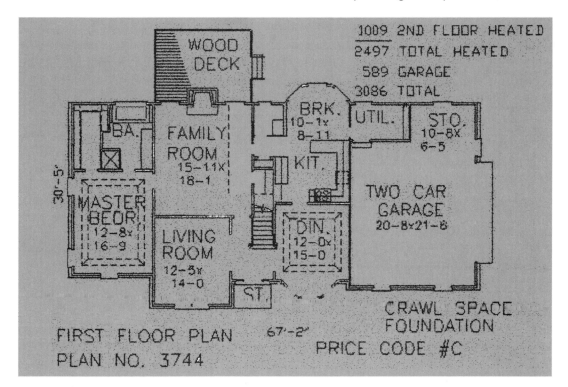

Figure 8.5 Standard house designs from the Internet.

The construction industry is likely to prefer closed networks set up for particular projects, but it will use the public networks to promote its services electronically.

8.9 Changing roles in construction

Computer systems, particularly those using knowledge-based techniques, enshrine the expertise of particular specialists. Construction applications have already led to work, traditionally carried out by one profession, being taken over by another. The demise of the bill of quantities has long been predicted but, if this ever happens, it is more likely to come from the need for simpler forms of description than as a by-product of a CAD system driven by an architect or engineer. Engineering skills involve rather greater liability than those of architects. The output of a structural design program has to carry the authority of the engineer who ran it and approved the results.

With architecture the options are much wider. Many house owners and business leaders believe they can design their own buildings. The skill of the architect in creating new opportunities and logical spaces is often not fully valued and, if the ability to draw, and even select components, is available to lay people, they may believe they can develop a design sufficiently to give drawings directly to a builder. There are already standard house plans available on the Internet (*Figure 8.5*).

Technology Foresight – Construction [4] asked the question: when would 'clients use workstations to access design and engineering services?'. The majority of respondents gave a date of 2004 for this but felt it would have little effect on wealth creation or the quality of life. Just as many who want to build their own house select plans from books or become self builders, so the tools for design and drawing are available to all through CAD systems.

General graphics and even specialist draughting software is becoming available at low cost and is likely to be bundled with office automation software in future.

8.10 Hybrid CD-ROM and Internet data

A characteristic of many of these future trends is the combination of human skills with those of IT, and the integration of different technologies. Most now realize that IT will not automate everything and that it is highly inferior to many human skills, that is until we can fully understand and emulate the brain. Trends in data retrieval (see Figure 6.6) have shown on-line data vying with that on fixed media such as CD-ROM as if they were directly opposing technologies. Each has its strengths and the combination of the two in the form of hybrid CD-ROMs is more powerful than either.

CD-ROMs are stable and cheap to produce. They, and the new DVD standard, are increasing capacities to several hours of video, thus meeting the needs of the entertainment industry. On-line data can be more current and allows communication and discussion, but it is not necessary for data which changes very slowly. The hybrid CD-ROM allows direct access to Web sites with the fastest changing information. *The Chronicle of the 20th Century* by Dorling Kindersley does just this with annual publication of a CD-ROM linked to a Web site containing the most recent data, which is then incorporated in the next issue.

More systems in future are likely to combine different technologies where each is deficient in some respect. Convergence will also be seen in linking computers to home entertainment centres and high bandwidth communications. This will bring on-line data into many homes, including shopping and referenda, and will be a major factor in creating the information society.

The EC RACE project BRICC and its successors on Collaborative Integrated Communications for Construction (CICC), Mobile Integrated Communications in Construction (MICC) and Reconstruction using Scanned Optical Laser and Video (RESOLV), have shown how technologies can be combined to capture multimedia data and transfer it from site to design office. Video data can be collected using an intelligent hard hat, transmitting it from the site to the office (*Figure 8.6*). Existing buildings can be modelled by means of a video survey trolley. Project networks display video images of the site, or the project team, to aid construction decisions.

Figure 8.6 Hard hat with communications and data display. BRICC project, BICC and others.

Both in the home and on the construction site, the combination of IT facilities in novel ways will provide new possibilities, particularly as a result of being able to transmit multimedia data rapidly and reliably.

8.11 Summary

A number of general trends will continue into the future: further miniaturization of processors through molecular level circuitry leading to even faster systems which can be carried on the person as hard hat or wrist-worn computers. Seamless communications will use broadband and broadcast networks to deliver live video images anywhere. This could lead to total surveillance which is not just for security but to link different parts of an organization or a virtual project team. Even friendlier software will use knowledge-based techniques to capture human experience and learning techniques to gain human skills and begin to make judgements.

New problems, and perhaps some of those that have been known for years, will be solved with hybrid technology using much higher density storage media and linking in-house and project Intranets to corporate object databases. The World Wide Web will carry the main sources of information, polarized into the better organized commercial sites and an ever growing body of shared and promotional data.

The great, but elusive, goal for building will continue to be the common building model and libraries of all the data needed to design and construct.

How this evolution will take place, and what conditions will need to change in construction, are the subjects of the next chapter.

9

Lessons for future technologies and their adoption in construction

9.1 *Characteristics of change*

Having examined some of the more detailed changes likely to take place in IT and their effects on construction, there are some general lessons which can be drawn on the characteristics of future change. There are few new inventions. Most developments are based upon ideas which have evolved gradually, even if specific products are claimed as new and particular individuals credited with their invention. Some of the most significant features of innovation are summarized here and the opportunities they provide for construction are discussed in sections 9.2–9.5. Their significance for the user of future technology is that, to commit resources and the time needed to learn new systems, requires some assurance of future success. Some of the factors which have proved significant in recently successful technologies have been identified, but there are other factors which are difficult to foresee. Any successful adoption of new technology requires some change in the user's organization.

Evolution – Technology advances by evolution and the first of a line of new products is often a failure and may be put aside until others have improved upon it, or related technologies, on which it is dependent, have advanced sufficiently. Unfortunately evolution in business is often regarded as plagiarism and developments derived from other firms' products rarely acknowledge the link and may make superficial changes to hide it. The terminology associated with software is one example, where new words are coined to describe the same processes, and this makes it very difficult for a user to change from one system to another. The British Standard on CAD layers [7] included a table of equivalent terms from different systems to help reduce the confusion. Without the threat of law suits, evolution could proceed in a more continuous manner and the contributions of all could be acknowledged.

Market share – The stage of its evolution when a product becomes established is usually when it achieves a significant market share. This may be quite a small proportion of the whole market where there are many competing products. For the technology as a whole to succeed, particularly if it involves communication, it must reach critical mass, usually regarded as about 30% of its market. In this case rival products help each other to reach this target together. In recent years particular groups of products, notably those of Microsoft, have achieved dominance through strategic alliances and providing access for developers of related software.

Hybrid technology – The vertical thinking approach to development of new products is to find one technology or product which meets the whole of the need itself. As the obvious IT products which can follow this approach have already been developed, the personal computer for example, lateral thinking is necessary to meet more complex needs. This may include combining technologies into hybrid systems such as information systems which provide links between fixed media, such as CD-ROM, and on-line information on the Internet. Other examples are the convergence of computing and communications in the form of the portable phone with a palm top computer, or the set-top box linking a TV to cable or satellite communications.

Timing of change – The launch of a new product must be timely to succeed. This implies timeliness, not only for the technology to be new yet sufficiently robust, but also for the market to be ready for it. The Apple Newton personal digital assistant was launched in 1993 as a hand-held machine accepting handwriting as input. It was not an immediate success since the market was still evolving from lap top to palm top computers and accepted keyboards as a necessary evil. The ability of people to adopt new technology develops more slowly than the technology itself, and organizations, such as those in construction, change even more slowly. A new product has to educate the market and, if it is to get the best from it, that market needs to change. Most IT products which meet specific and distinct current needs have already been established.

Changes needed in construction – The man–machine interface has evolved to a very sophisticated level yet the industry–machine interface is relatively primitive. IT takes as its primary market the human being and can be developed ergonomically to suit the needs of human hand, eye and brain. Its intelligence is modelled on what we know of human intelligence through neural networks and virtual reality, in its total immersion form, and aims to stimulate sensory perception. Groups of people, professions or industries, are not well enough defined to be simulated so easily. Networks and communication systems are the key to representing flows of information in construction and are needed to connect clients, designers, building sites and contractors. A nervous system is needed for each project. This should convey information to all involved, feel pain where there is a problem, and convey messages from its eyes as well as directly to its brain.

The essence of the industry–machine interface is the combination of different types of data in multimedia form, the communication of this data, the addition of human experience through greater intelligence, and portability of the devices to deliver it.

Even with all these developments, which the previous chapters have described, well under way, they will only thrive in an industry which is well organized and can provide good data in the right form.

9.2 Evolution in construction

Design is a cyclical process, as any architect will maintain to a client or fellow consultant. The first cycle is often driven by an idea, possibly to the exclusion of practical considerations. The second is modified to meet pragmatic needs, and the third often refines these into an acceptable solution. In IT the third cycle in the development of a product is frequently the best, and further refinements often add unnecessary features which can make the product excessively large and complex.

> Windows 3 was the first release which really worked although it was actually version 3.1 which established Microsoft's graphical interface. The Apple 3 was not a success but Apple 1 was a very preliminary prototype. The Macintosh could be said to be the third release by Apple computer and introduced the graphical user interface. In CAD there were mainframe systems and minicomputer systems but usage only spread when CAD became available on microcomputers. Autocad version 1 was very primitive; it improved with version 2 in 1984 and version 2.5 in 1986. It then became so successful that the third version was launched as Release 10 in 1988. Data exchange between CAD systems has been through two formal stages of definition: IGES was never taken up in construction, STEP is still being defined, while the DXF format which everyone was using emerged from the grass roots backed by the market dominance of Autocad. Its success, and the more ambitious goals of STEP, will be continued in the Industry Foundation Class libraries being produced by the IAI.

Construction is evolving in parallel with, if rather more slowly than, information technology. It is probable that IT is now starting to catalyse some of this change although the most radical changes of recent years have been driven by commercial necessity and by the Latham report's exhortation towards a more harmonious industry. This harmony is a prerequisite of using IT to its full advantage. There must be a meeting of minds by all those involved in construction before there can be free and efficient exchange of data. Joint education of the construction professions is a part of this and a sensible sharing of information will extend this into projects for which the expertise of several specialists is essential.

Data has been used as a defensive wall between participants in the building process. The minimum necessary to carry out the project should be supplied, preferably in a form defined by the recipient. The tendency in the past has been for each participant to produce excess information to overwhelm the rest of the project team. This approach can no longer be afforded and an idea, which will work with the concept of partnering, is that the minimum necessary information is held centrally and each participant contributes to, and takes from, this central resource the concise data they need. The British Property Federation proposed, in their form of contract, that payment was

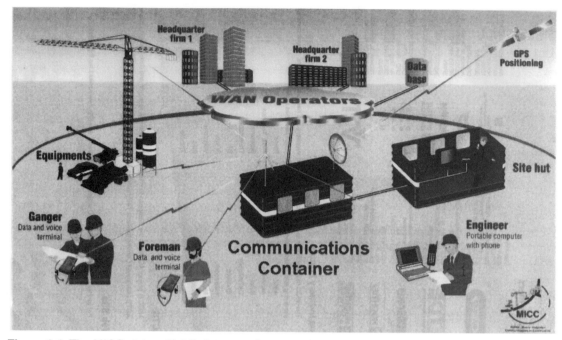

Figure 9.1 The MICC vision. Mobile integrated communications for construction. A European Commission research and development project. BICC and others.

made on delivery of specific items of information – typically drawings. IT implies that deliverables must be defined, not as specific documents, but as the content of the project model to a certain level of detail. The principle of payment for delivery of minimum agreed levels of information can still apply.

The growth of construction networks, the essential basis for better communications between the many groups involved, requires several types of system: industry information, project information and links between all participants and with building sites such as those proposed in the EC MICC project (*Figure 9.1*). Background information on legislation, standards and products is already available on-line and on disc but, as use of the Internet and optical disc develops, the structure currently imposed by the small number of information intermediaries could be lost if those who own the data decide to supply it direct in their own format. There is also the danger that information produced, originally with public resources, will become too highly priced for general use.

A Freedom of Information Act would allow low cost access to a vast body of data already collected where it has not already been dispersed into private enterprises.

Project information networks are likely to be the most productive in terms of improved service to the client. Should they be set up for each project or form part of larger, existing networks? This will depend upon the success of partnering and the willingness of individual firms to provide access to their computers. When such networks are in place, the participants could be located anywhere in the world and the idea of the virtual project team becomes feasible.

Links to sites will be part of the contractor's operation but progress on the project is of interest to all those involved and, with greater use of surveillance cameras in society at large, views of the site, whether general ones to show progress or more specific shots directed by site staff to solve problems, will become standard practice. The BRICC project on Broadband Integrated Communications for Construction [52] and its successors have shown how information can now be passed effectively from site to office and between the members of the project team. As building models evolve to simulate the construction process, so a library of recorded video material of previous uses of products or typical projects, could be built up. The day of the site operative with a visual display attached to a helmet showing images of drawings, people and procedures, may not be far off (see *Figure 8.6*).

9.3 *The construction market*

The construction industry in the UK represents 7–8% of GDP and, if all the groups within it pulled together, it could have a significant influence on IT products and services. It represents a lower percentage of GDP than most of its partners in Europe and is therefore likely to grow as harmonization proceeds. Its needs in general administrative and management areas are sufficiently close to those of other industries to help create the critical mass for office automation and communications. It is in those areas where it has special needs that consultants and contractors should be co-ordinating their systems. Industry information, organized in standard ways and accessible through a small number of gateways, is needed to provide the basis on which individual projects can build. Communications networks, where those internal to companies can be linked to others worldwide, perhaps via Intranets and the Internet, are the means of building virtual project teams. Design and analysis around an intelligent building model, which is maintained from concept to demolition, is the key to an efficient process.

All these developments are under way – the technology exists – but it will only thrive in a framework of common understanding resulting from joint education, compatible systems, where marketing pressure to be different for its own sake is suppressed, and where accessible models are owned and maintained by the project.

The nature of firms in construction is already changing with some influence from the opportunities provided by IT. Large firms are generally getting larger and more multi-skilled even if their specialisms are held in subsidiary companies. The Private Finance Initiative means that they will have to become more involved in both the financing of the project and the management of the resulting building, areas in which IT has not been used as fully as it might. Partnering will set up longer term relationships in which new communications will thrive. Smaller firms need not necessarily suffer since they can offer specialist skills or local knowledge via the public networks which, although they may become more expensive to use, offer access from anywhere. It is only the smaller firms ignoring the technology which may suffer, but there will still be local needs only they may be able to satisfy.

> An RIBA IT survey in 1980 [59] found that 19% of architects said they did not use computers and would never see a need to do so. The following survey in 1994 found this number had fallen to 10.8%, mostly single practitioners [64]. IT has now become just another aid to practice. It is getting into almost all offices in the form of word processors or draughting systems. One architect still holding out against this technology is Quinlan Terry whose work is largely based on Palladian style buildings. He believes computers are for the workshy and that they provide opportunities to play for people who are bored [60]. This approach should be respected but a more constructive view is that IT is just another aid to business and need not affect the design of the buildings which it can help to produce. In time the discrete processes to which IT has been applied, will merge into a total flow of data through the project, which larger firms and projects are already starting to employ.

9.4 Hybrid technology

In construction the traditional groupings of professions and companies are breaking down with project teams being assembled from new combinations of skills. This is in response to changing demands, the greater involvement of clients and finance, and the need for more effective management. IT has supplied the most obvious applications for construction, the software which automates particular tasks, and new solutions are needed for integration, increased intelligence and to meet the demands of these new types of project team.

Hybrid technology is often the route to meeting new needs. Marketing tends to polarize products into rival camps while the users' needs may lie in breaking down the barriers between them. One of the dilemmas in choosing desk top computers during the 1990s has been between Macintosh and Windows PC, yet the Power PC processor was developed by Apple, IBM and Motorola to make these technologies converge. Unfortunately the Intel family of chips has prevailed and the only hope for keeping the Macintosh in the mainstream is the opening up of its excellent technology to licensees or the use of dual processors.

Another apparent conflict is between the optical disc and on-line systems as the means of retrieving information. Each has its merits and the hybrid solution is to keep the constant data on optical disc with pointers to World Wide Web sites, which can be accessed directly from the disc via a modem, and will present more recent data in compatible form (*Figure 9.2*). Another link is between software and disc based information, for example that between Autocad and the Barbour Construction Expert, a CD-ROM based aid to selection of building products. Encyclopaedias, like Encarta, have an update button which links the user to the latest on-line information. This is then incorporated in the next issue of the CD-ROM. A quarter of all CD or DVD-ROMs are expected to be hybrid ones by 1998 [61].

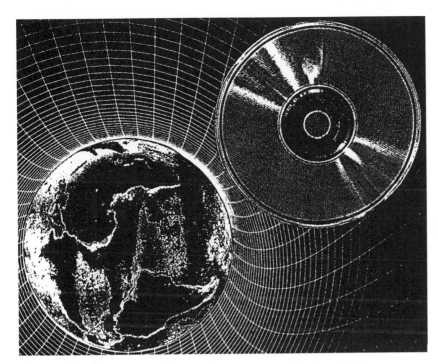

Figure 9.2 Hybrid information sources. An image of CD-ROM and the Internet.

Another current debate in IT is between network computers with no local storage, and the Windows PC, which holds all its own software and data. There will be examples of each type suited to different sizes of organization, but PCs will have good communications when they are needed and users of NCs will inevitably want some local storage and the ability to use regular applications such as word processors and spreadsheets independently of the networks. The operating systems debate between UNIX and Windows is likely to converge on Windows NT which can run on all the main types of computer and provides for multiple users and multi-processing as well as having a friendly and familiar interface. The competition will come from Web browsers and languages, such as Java and ActiveX, which will become ever more familiar as more transactions are conducted over the Internet.

Office machinery has tended to be dedicated to particular tasks but several types of equipment share the same printing technology. Hybrid office machines are now being developed, particularly for those working from home. In 1997 Hewlett-Packard announced the OfficeJet Pro a combined colour inkjet printer, colour copier and flatbed scanner. This handles A4 paper and costs under £700 with a speed up to 5 pages per minute and resolution up to 600 × 600 dots per inch.

In applications software, the hope for a single master database linking them all has given way to a more incremental approach with applications, commonly used together, being linked by a shared database. Examples are: CAD and quantities, lighting and visualization, environmental analysis and services design, project management and resource control. The addition of a database provides new opportunities for managing data. CAD systems are moving towards the management, not just of drawings, but related project documents and their storage and distribution. Autodesk provide their Workcentre product for aiding design team workflow, and Bentley Systems have Backoffice and Teammate. Work flow addresses the whole design and documentation process of which drawing production is a small part.

Databases lie behind Document Management and Geographical Information Systems, and data may be captured for these by scanning, or from digital maps, and its manipulation can be performed by database management software. Facilities Management provides similar tools for managing buildings and their contents and services.

Before seeking entirely new software to meet fresh needs, the combination of existing solutions should be explored since, where these already have databases, the linking of these should be possible. It depends on the structure of the data and this is now the critical element in future integration of applications for construction.

9.5 Speed of change

The possibilities for change offered by IT can be seen long before the technology is reliable or the construction industry is ready to make effective use of it. The history of computing is littered with developments which took years to become effective or may still be waiting for this to happen.

Automated design was an early hope which never materialized, yet it remains a fascination and new techniques, such as neural networks and expert systems, may add the subtleties which were missing in the 1970s.

The single project database has long been a goal and is still being strived for, but the route to its eventual attainment is more likely to be by aggregation of individual components – good product data with the most widely used applications – rather than by going for the whole database in one go.

Voice technology is taking its time to become established. Anything less than reliable analysis of continuous speech has only a limited application, and continuous speech and automatic translation remain very difficult problems.

Virtual reality was first demonstrated in the 1960s but the total immersion variety is still very limited, far from reliable, and mostly used in games arcades. The on-screen, and curved screen projection versions are established but VR is really nothing more than another step in the evolution of CAD and simulation.

On-line databases and electronic mail have recently become widely taken up after being available for twenty years. The many small and technically different systems, charging by the time used, failed to convince construction users and it took the network of networks, which is the Internet, with its relatively modest fixed charges, to establish large scale usage.

The take up of these technologies by construction has often followed that by other industries. Individual firms have been pioneers in solving a particular need, or have wanted to be seen as advanced technologically. Changes in the construction process have rarely led to technological innovation although competitiveness and partnering are starting to encourage it. The stages for introduction of a new technology are first, research, either by universities or by IT suppliers, then adoption by leading firms with publication of their experience and, finally, a more general acceptance. For those firms which do not wish to be pioneers, it is best to wait for others' experience to be published and for the third release of software, typically the stage at which most of the problems have been solved. After that users' demands for additional, but inessential, features may lead to the software becoming more complex than necessary.

Occasionally a new technology appears relatively quickly and the elapsed time from research to reliable use can be quite short. In my 25 years' experience of IT the most exciting developments have included:

Storage tube displays – allowed the first display of vector graphics linked to timesharing computers using white lines on green screens, in about 1970.

Spreadsheet software – a general capability of computers for tabular calculations which first became widely available as Visicalc on microcomputers – 1979.

The standard PC – for the first time software became widely interchangeable between different computers and the tendency of hardware suppliers to lock users into their systems was broken.

Low cost CAD – from being highly expensive and stressful systems for use by specialist staff, computer-aided design became another low cost application accessible to all.

Multimedia – arrived with the convergence of all forms of data in digital form, and the availability of PCs with CD-ROM drives, sound cards, modems and colour screens to handle this, in the early 1990s.

The Internet – another long evolution in the academic world which then burst upon the business community to provide cheap communications in about 1994.

It is interesting, if dangerous, to speculate on which developments heralded for the future might generate similar excitement:

Integration of computing and communications – the palm top computer with mobile phone will create as much change as the portable phone. Also set-top boxes linked to a digital TV, will penetrate the home to provide entertainment and shopping.

Full immersion virtual reality – when this works effectively and, when fully rendered models can be explored interactively, this will use helmet displays and sensitive body suits to communicate with our senses and create a truly artificial environment.

Electronic commerce – the arrival of valuable data on the Internet with a modest payment for access to particular pages and invoicing through telecoms providers.

A Freedom of Information Act – would provide access to data generated for the public while it is still owned by the state. Digital maps are one example of data which is very expensive while in the USA such Federal data is available at low cost.

Reduction of travel to work – effective communications allow more people to work from remote locations for part or all of the time. This was given a high priority by those responding to the Technology Foresight study [4] and would increase the urgency, for construction, of developing the concept of the virtual project team.

A common framework for construction data – could evolve from the experience of large firms with access to public data at low cost, recognized by bodies such as the Construction Industry Board, which is responsible for unifying the construction industry.

Some of these developments have been evolving for many years but, when and if, they are finally established it will be an exciting time for all those in construction.

9.6 Changes needed in construction

After nearly forty years of information technology in construction, the systems available to us have adapted themselves quite effectively to the individual user. They have yet to reflect the needs of groups of users, particularly in such complex and changeable forms as those in construction. IT research shows promise in the way that computers will relate to the body and feed data to its senses. Research is also revealing more of the working of the brain, and IT systems will be able to emulate some of this. They are already capable of transmitting data across the world. However, organizations are not as easy to study as the human body. Human organs are rarely in competition with each other and our nervous systems communicate very effectively.

It would be nice to think that industries can be understood and organized in a way that allows IT to demonstrate its full potential. There are examples in countries smaller than the UK, in Scandinavia for example, of well-organized construction data, and, in some projects, of truly collaborative partnering. This is unlikely to spread throughout all construction projects and IT systems must continue to respond to the varying needs of an industry which will continue to be diverse and complex.

The facilities offered by IT are starting to influence the structure of the industry and new services are being established. These include networks linking groups of firms and carrying project data, construction information sources on fixed media, tools for collaborative group working and transmission systems for any sort of data. The problem is that such systems proliferate and new technology is often driven by marketing departments promoting their products and services as different (and better) than their rivals'. What is needed is a mechanism to recognize the best, or most successful, of these and reinforce their position as market leaders. Becoming established as market leaders, as Autodesk and Microsoft already have, can be resented by competitors but, once their status is recognized, they give shape to the market, set *de facto* standards, and their rivals can still claim to have superior products.

> Strategic alliances are often the best means of reinforcing success and an example has been set by Autodesk in its Collaborative Design and Engineering strategy which aims to implement some of the recommendations of the Latham report (*Figure 9.3*). This is one of the first initiatives to address the more complex needs of construction. In addition to linking the various levels of CAD software in AutoCAD, it also links to other successful products such as: the National Building Specification, document management, facilities management, the Barbour Index Construction Expert and the Internet. Such alliances are important and should be recognized as giving a diverse industry some structure, however er changeable this may be as its constituent parts evolve.

Given a dominant product such as Microsoft Office, related software will be provided with a means of exchanging data, via OLE for example. The danger of too much dominance by a single supplier is that complacency may set in but, if there are linked and competing products, these keep the leaders on their toes. This book reflects some of the products dominant in the mid-1990s and gives pointers to the technologies, if not the specific products, which are most likely to succeed in future.

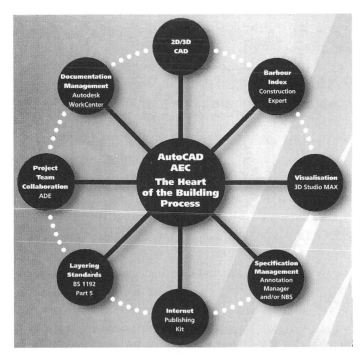

Figure 9.3 Autodesk collaborative design and engineering strategy. An example of new forms of integration with data linked to their core CAD product, AutoCAD AEC.

Construction has a tendency to want to keep its options open and to design every new building as if it were the first. This is a result of competitiveness and the higher profit margins enjoyed in the past. In future it could benefit from a little more conformance and a willingness to reuse methods, components and data. For routine tasks there is no need to look for novel solutions. For work which requires special attention, to meet the needs of special projects or to provide 'commodity, firmness and delight', the imagination which exists in construction should not be constrained.

IT systems are constantly changing and offering new opportunities but they are merely an aid to an information transfer process which, if a little better organized, could change the way in which the client's needs are communicated right through the design and construction process. Good building data is becoming the most important element in IT systems and, for a completed project, this should be returned in the form of an electronic tool for managing a building over its lifetime.

9.7 Making the most of IT

In trying to assess future technologies, it should be remembered that information systems are the servant not the master. They may have a strong influence on our culture, the way construction businesses are run, or the form which buildings take, but other factors are more important. Over the time in which I have been involved with computers they have evolved from being a fascinating and remote possibility to an essential element of any business.

The general trends are clearer than the future of specific products. In time all published information will be available in electronic form. The devices to access it will be small, cheap and universal, communicating with most of our senses and even directly with our brains. Software will have much greater intelligence and will explain its reasoning better to reassure reluctant users. Communications will be seamless, from the domestic TV to the information resources controlled by media moguls around the world. Commerce and entertainment will be delivered rapidly, when and where it is required by its consumers.

The construction industry provides a very large market for IT and has very much to gain from better communications. It is used to networking, in its social sense, to obtain business and exchange experience. It now needs to implant its own structure, on a global scale, on information networks and set up project networks for managing the explosion of data which will occur.

The electronic persona of the industry, its various parts and its many individuals, will need to be armed with knowledge agents to help them find good industry data and manage project data. Media moguls, information providers and competitors from other countries will have their own agents trying to introduce their own data.

The evolution of IT in construction in the UK should build on its long experience and anticipate and adopt the technologies which will serve its future interests best.

References

1   Negroponte, Nicholas (1970). *The Architecture Machine*. MIT Press.
2   Basalla, George (1988). *The Evolution of Technology*. Cambridge history of science series.
3   Mitchell, William J. (1996). *City of Bits*. MIT Press.
4   Office of Science & Technology (1995). *Technology Foresight – Construction*. HMSO.
5   Construction IT Forum (1990). Building IT 2000. (1995) Building IT 2005. Glaxo Research Campus.
6   Evans, Christopher (1980). *The Making of the Micro*. Gollancz.
7   British Standard (1990). BS 1192 Part 5. Construction drawing practice. Structuring of computer graphic information. BSI.
8   Butler, Samuel (1863). *Darwin Among the Machines*.
9   *Architectural Forum* (1943). Building for 194X.
10   *Byte* (September 1995). 20th anniversary special issue.
11   Draft British Standard (1997). Revision of BS 1192 Part 5. BSI.
12   Draft International Standard (1997). ISO 13567 Technical product documentation – organisation and naming of layers for CAD. ISO.
13   Negroponte, Nicholas (1995). *Being Digital*. Hodder & Stoughton.
14   Giedion, S. (1967). *Space, Time and Architecture*. Oxford University Press.
15   Brino, Giovanni. (1995). *The Crystal Palace*. Sage.
16   Alexander, Christopher (1964). *Notes on the Synthesis of Form*. Harvard.
17   Rudofsky, B. (1964). *Architecture without Architects*. New York, Museum of modern art.
18   *Chartered Surveyor* (October 1984). *Building* (February. 1965).
19   Exhibition (1968). *Cybernetic Serendipity*. Institute of Contemporary Arts.
20   Conference (1972). *Computers in Architecture*. York. RIBA.
21   Auger, Boyd (1972). *The Architect and the Computer*. Pall Mall.
22   Survey (1973). *Computer Aided Building – a Study of Current Trends*. ARC and SPL for Department of Trade & Industry.
23   Report (1978). *Final report on computing and data processing in the construction industry*. National Consultative Council, DOE.

24   Latham, Sir Michael (1994). *Constructing the Team*. HMSO.
25   Report (1988). *Building Britain 2001*. Centre for Strategic Studies in Construction, University of Reading.
26   *Building* (10 January 1997). Economic outlook.
27   Atkins Consultants (1993). *SECTEUR Strategies for the European Construction Sector – a Programme for Change*. European Commission.
28   Guidance Note No 36 (June 1992). *Contract Strategy for Major Projects*. CUP.
29   *Building* (September 1996). List of leading clients.
30   Day, Alan and Langford, Victoria (1989). *The Implementation of Coordinated Project Information in the UK Building Industry*. University of Bath.
31   Current paper (1981). M/81 *Quality Control on Building Sites*. Building Research Establishment.
32   Cross, Nigel (1977). *The Automated Architect*. Pion.
33   Survey (1993). *Building on IT for Quality*. CICA, KPMG.
34   *Architects Journal* (January 1983). 'Buildings designed with computers.'
35   Benchmarking best practice (1996). *Construction Site Processes*. Salford University, Construct IT Centre of Excellence.
36   Benchmarking best practice (1997). *Briefing and Design*. Salford University.
37   Benchmarking best practice (1997). *Facilities Management*. Salford University.
38   Toffler, Alvin (1970). *Future Shock*. Pan.
39   Kinsman, Francis (1980). *Millennium*. W H Allen.
40   Annual surveys (1981–1995). *CAD Sales Surveys*. CICA September Bulletins.
41   Betts, M. and Oliver, S. (January 1996). An IT forecast for the architectural profession in *Automation in Construction*. Elsevier.
42   World Wide Web site (1997). Internet Society. http://info.isoc.ortio.
43   Report (1995). *Construct IT – bridging the gap*. Andersen Consulting for DOE.
44   Annual report (1996). *Software Directory*. CICA.
45   World Wide Web site (1996). *Construction Industry Gateway*. www.bre.co.uk/~itra/cig/
46   Richens, Paul (1990). *Microcad Software Evaluated*. CICA.
47   Multi-client study (1990). *The UK AutoCAD Market*. Cambashi.
48   *Management Today* (1996).
49   *The Observer* (5 January 1997). The search for net profits.
50   Report (1995). *Quantifying the Benefits of IT*. CIRIA/CICA.
51   CD-ROM. (1990–1996) Byte on CD-ROM. *Byte*.
52   World Wide Web site (1996). BRICC project. BICC www.hhdc.bicc.com/bricc.htm.
53   *RIBA Connect* (November 1996). Offices for Andersen consulting.
54   Taylor, David A. (1990). *Object Oriented Technology, a Manager's Guide*. Addison Wesley.
55   Technical Report 14177 (1993). *Classification of Information for Construction*. ISO.

56   World Wide Web site (1997). Cyberatlas – Electronic commerce. Forrester Research. www.cyberatlas.com/emoney.html.

57   US Government (1996). Report on electronic commerce.

58   *Byte* (September 1996). Activmedia. Survey.

59   *RIBA* (1980). Computer use by architects. Survey.

60   *RIBA Connect* (February 1997). Paper chase.

61   *Personal Computer World* (February 1997). Hybrid CDs.

62   Market report (1997). *The Barbour Report*. Barbour Index.

63   Gates, Bill (1996). *The Road Ahead*. Penguin.

64   RIBA (February 1996). Computer use by architects. Survey.

65   *Oxford Dictionary of Computing* (1996). Oxford University Press.

66   Report for EC DG XIII (1982). The specification of a building industry workstation. CICA, I3P and the Technical University of Munich.

67   *Building* (April 1967). Computer conference.

Glossary

ACADS	Association for Computer Aided Design – former Australian user association
AEC	Architect, Engineer, Constructor
Analogue	Relating to constantly changing physical values
ANSI	American National Standards Institute
APEC	Automated Procedures for Engineering Calculation
Applets	A miniature application typically built into Web languages, e.g. Java
ARC	Applied Research of Cambridge
ASCII	American Standard Code for Information Interchange
ATM	Asynchronous Transfer Mode
BAA	Formerly British Airports Authority
Bandwidth	The range of frequencies that can be passed by a transmission channel
Bit	Binary digit – the basic unit of data represented as either 1 or 0
BRE	Building Research Establishment
BRICC	Broadband Integrated Communications in Construction
BSI	British Standards Institution
CAD	Computer Aided Design (includes draughting and modelling)
CCITT	Comite Consultatif International Telephon et Telegraph
CD-ROM	Compact Disc – Read Only Memory
CEN	Comité Européen de Normalisation – EU standards body
CEPA	Computers in Engineering, Planning & Architecture – former US user association
CIAD	Computers in Architecture and Design – former Dutch user association
CIB	International Council for Building Research Studies & Documentation

CIB	Construction Industry Board
CIBSE	Chartered Institution of Building Services Engineers
CICA	Construction Industry Computing Association – UK user association
CIOB	Chartered Institute of Building
CI/SfB	Classification system originating in Sweden
CLASP	Consortium of Local Authorities Schools Programme
CPI	Co-ordinated Project Information
CP/M	Control Program/Microprocessors
CRISP	Construction Research & Innovation Strategy Panel
CSTB	Centre Scientifique et Technique du Batiment
DEC	Digital Equipment Corporation
De facto	By virtue of things as they are
Delphi	Form of survey with multi stage answers based on previous responses
Digital	Relating to the storage and transmission of data using digits
Digitiser	A device that converts data into digital form
DOC	Design Office Consortium, later became CICA
DVD	Digital Video Disc
DXF	Data Exchange Format
EDI	Electronic Data Interchange
E-mail	Electronic mail
Express	A data modelling language commonly used for product models
FACE	International Federation of Associations of Computer Users in Engineering, Architecture and related fields
FMB	Federation of Master Builders
GDP	Gross Domestic Product
Hardware	The electrical and mechanical parts of a computer system
HTML	Hypertext Mark-up Language
IAI	International Alliance for Interoperability
IFC	Industry Foundation Classes
IGES	Initial Graphics Exchange Specification
Internet	Many small networks which are connected to cover the world
Interoperability	Co-operative sharing of data between different programs
Intranet	Local area network using similar technology to the Internet
ISDN	Integrated Services Digital Network

ISO	International Standards Organization
Isometric	Three-dimensional projection with the same scale on all three axes
IT	Information Technology
JCT	Joint Contracts Tribunal
Kbytes	Thousand characters of information (1024 to be precise)
Knowledge agent	Software which will search networks to find specified data
LAN	Local Area Network
Lap top	A portable computer that can use batteries
Macintosh	Computer made by Apple with early graphical user interface
Mainframe	A powerful computer for processing complex data, often at the centre of a network
Mbytes	Million characters of information
Microcomputer	A small digital computer containing one or more microprocessors
Minicomputer	A computer that is smaller and slower than a mainframe but larger and faster than a microcomputer
MIT	Massachusetts Institute of Technology
Model	A simulation which describes how a system behaves so that a program can control the system or explore the effects of changes to the system
Modem	Modulator/demodulator for linking computers to telecom lines
MSDOS	Microsoft Disk Operating System, an operating system for personal computers
Multimedia	The combination of sound, graphics and video to present information on a computer
NEDO	National Economic Development Office
Neural networks	A method of computing in which elements receive data and links are made as repeated patterns are recognized
Object classes	Software packets containing related data and methods which are identical in form and behaviour but contain different data.
OLE	Object Linking and Embedding
OpenDoc	An alternative compound document specification similar to OLE but with many more advanced features
Optical fibre	Cable of glass or plastic which carries signals of light modulated to transmit data
Orthogonal	Right angled

OSI	Open Systems Interconnection
Palm top	A smaller version of a laptop computer which can be held in one hand
Pascal	A high level computer programming language used in the development of microprocessors
PC	Personal Computer, originally compatible with the IBM personal computer
Pixel	The smallest single point on a visual display screen
Plasma	A flat screen with gas between layers of glass in which characters are formed
PSA	Property Services Agency
RIBA	Royal Institute of British Architects
RICS	Royal Institution of Chartered Surveyors
SCOLA	Schools Consortium of Local Authorities
S-curve	Typical growth pattern for communicating technologies
SEAC	South East Architects Consortium
SMM	Standard Method of Measurement
Software	Programs that run on a computer
STEP	Standard for the Exchange of Product model data
Storage tube	Graphical display in which lines are retained on the screen
TCP/IP	Transmission Control Protocol/Internet Protocol
TFT	Thin Film Transistor technique for flat display screens
Timesharing	The use of a computer system by more than one user at the same time
Uniclass	Unified Classification for the Construction Industry
UNIX	An operating system for different computers from mainframes to PCs
Videotex	A system displaying information on a specially adapted TV screen
Virtual reality	An environment created in a computer and allowing interaction in real time
Web	World Wide Web – a hypermedia system including graphics on the Internet
Windows	A Microsoft operating system showing data in different areas on a screen
Wireline	A way of representing 3D objects on a screen by only drawing their edges

Index of people and topics